George Henry Kinahan

Valleys and Their Relation to Fissures, Fractures, and Faults

George Henry Kinahan

Valleys and Their Relation to Fissures, Fractures, and Faults

ISBN/EAN: 9783744742931

Printed in Europe, USA, Canada, Australia, Japan

Cover: Foto ©berggeist007 / pixelio.de

More available books at **www.hansebooks.com**

VALLEYS.

PRINTED BY BALLANTYNE AND COMPANY
EDINBURGH AND LONDON

Frontispiece.

COOMS AND OLD SEA CLIFF IN THE EAST SLOPES OF DERRYCLARE AND BENNABEOLA.

VALLEYS

AND THEIR RELATION TO

FISSURES, FRACTURES, AND FAULTS.

BY

G. H. KINAHAN, M.R.I.A., F.R.G.S.I., &c.

OF H.M. GEOLOGICAL SURVEY.

LONDON:

TRÜBNER & CO., 57 & 59 LUDGATE HILL.

1875.

HIS GRACE THE DUKE OF ARGYLL,

K.T., P.C., D.C.L., F.R.S., &c. &c. &c.

My Lord Duke,

Your Grace's contributions to
Physical Geology, and your able Presidential Addresses to
the Geological Society, London, have clearly manifested
that, whilst giving full weight to those arguments on Subærial Denudation which have of late years been so domi-
nating with the modern school of Geologists, you have
not permitted them to exercise an undue influence over
your conclusions, but have been enabled to trace to other
and yet earlier geological agencies, also, the physical features of our Earth's surface.

It is therefore with no ordinary satisfaction that I avail
myself of your Grace's permission to dedicate to you this
brief attempt to maintain, by observations in the field,
those views so ably advocated by your Grace.

I have the honour to be,

My Lord Duke,

Your Grace's most obliged,

and most obedient Servant,

G. HENRY KINAHAN.

H.M. Geological Survey,
Wexford.

LIST OF ILLUSTRATIONS.

CONTENTS.

PREFACE.

THIS book was commenced some years ago, and was to have been a joint production of my colleague Mr J. L. Warren and myself. His lamented and early death, due principally to his over-zeal in the discharge of his public duties, prevented the original plan from being carried out; it therefore devolved on me to complete it alone, and since then much matter has been collected, with which I have incorporated papers that have appeared in different scientific magazines and publications, on branches of the subject, but at the same time the original style has been adhered to, as if the joint-authorship still continued. This has been found necessary, as otherwise the credit for the matter and facts collected by Mr Warren, would be appropriated by me.

Some few suggestions that may appear in this book might, perhaps, not be sanctioned by that observer; these, however, can be but few, as most of the conclusions were discussed by us together, and the constant result of our discussions was agreement. Any facts which are stated without mention of any authority were observed by one or other or both of us; while facts, the knowledge of which is due to other observers, are given under their authority.

This book is, in the first place, a record of observed facts, but the conclusions we have arrived at may be controverted by others; however, we have tried to avoid putting forward anything that is not in accordance with precedent and ascertained operations of Nature.

Plates Nos. I. and II. were drawn by the late Mr Warren, while I am indebted to my colleague Mr M'Henry for Nos. III. and IV., and to Mr Nolan for the characteristic sketch of the gravel terraces in the Erriff river valley.

G. HENRY KINAHAN.

DUBLIN, *March* 1874.

INTRODUCTION.

PHYSICAL geologists are at variance as to the forces which have formed the present features of the earth's surface. It seems to have been formerly a very general belief that the shapes of the ground are more or less due to movements in the crust of the earth which have opened cracks and fissures, elevating some portions and depressing others, and that to these movements, combined with marine denudation, the features of the surface of the earth are due. Hutton seems to have been the first to broach the theory that the surface has been principally sculptured by meteoric abrasion; and now most of the working geologists in Great Britain seem inclined to disregard unduly every other kind of action. During many years of geological research, we have noted certain relations between the different kinds of " shrinkage fissures," breaks, lake-basins, valleys, and other features of the earth; and although formerly we accepted in a great measure the theories of the " subaerialists," we are now led to believe that the different *denudants* must act in combination;—each separately being incapable of doing much work,

A

while not one of the denudants could act efficiently
without the aid of external or internal *heat* to dry
and contract the rocks; because to the drying and
contraction of the surface rocks is due the minor
cracks and joints; and only in conjunction with these
shrinkage-fissures can any denudant work with effect.
Moreover, we are forced to believe that the faults and
breaks, due to the shrinkage of the earth's crust, have
materially assisted in giving the earth its present surface conformation.　Few geologists in these countries
now give any credit to faults, joints, and the like, as
having assisted in forming valleys or lake-basins,
while none give them the prominence to which they
appear entitled; but on the contrary, it has been
stated, "there is no necessary connection between
fractures and the formation of valleys."　It has also
been stated, that in many cases the materials "washed
out of the different valleys could not have been originally softer than the materials of the intervening ridges,
as the corresponding strata on the opposite sides of the
valleys are of equal hardness."　The latter statement,
however, quite ignores the fact, that if portions of the
ground are broken up and loosened in lines, by systems
of faults or joints, the intervening parts may be still
firm and unbroken; and when the shattered parts
are denuded away, the others would remain standing
in unbroken ridges and masses.

If meteoric abrasion is the universal denudant

some would wish us to believe, how is it that it has accomplished in these countries so little work since the glacial period, while the effect of marine denudation in the rearranged drift is so apparent? Many rocks, exposed ever since the glacial period to meteoric action, have not as yet lost their glacial dressing or striæ; while very little of the drift in the valleys above marine influences has been removed, and in other places no traces of this continuous denudation can be detected.

In the Esker gravels of the central plain of Ireland, but especially in the King's County, there are deep bowl-shaped hollows. The vegetable soil or decomposed drift on these gravels is always shallow, often only a few inches in depth. This, in ordinary cases, might be accounted for by its being washed away nearly as fast as it was formed, by rain and runlets, the slopes of these hills being always more or less steep. If such was the case in the bowl-shaped hollows, there ought to be at the bottom much more soil than on the slopes, the continual waste of the land carrying down the soil to the lowest level; yet such does not appear to be the case, the depth of the soil on the slopes and on the bottom of the bowls being nearly equal, except when the hollows are tilled. In these hollows, on account of the nature of the drift, water rarely, if ever, lodges. In the County Wexford much of the low country is covered by a drift composed of either marl or clay, in which cup-shaped hollows are not uncommon. The King's County bowls seem altogether due to eddies, or the piling up of the drift by transverse currents; while the Wexford cups seem to have been caused by a subsidence in the underlying strata. At first, probably, there were subterranean passages out of the latter; but now these are almost entirely, if not totally, closed, and there is now always more or less water in them, in which aquatic plants grow, which eventually produce a peaty accumulation. These accumulations, if a constant denudation were going on from the surrounding slopes, ought to be more or less earthy, or ought to have layers and thin seams of marl or clay in them, marking each summer's growth of the peat. This, however, is not the case, even when the adjoining land is in tillage; but on the contrary, as where sections can be examined, it is found that the peat tends to creep up over the drift.

Two or more systems of joints cross one another more usually at acute than at right angles; and we find, on looking at the map of the world, that the land is more inclined to form triangles than any other kind of figure. The tributaries of rivers or streams on sloping ground generally join them at acute angles, rarely at right angles. It may be asked, If the courses of streams are mainly due to joint-lines and such like, why then are there not networks of valleys, as each joint-line ought to be more or less developed into a surface feature in its whole length, and not only in part thereof? This, however, is easily answered, as water flowing down a slope denudes the weakest place, which must necessarily be a joint-line or some such crack, and eventually forms a channel along it. To this channel it is confined till it meets another channel, cut by a superior stream, into which it naturally flows, and loses all power of continuing its original course. This, however, is not always the case, as a stream coming down a joint-line belonging to one system

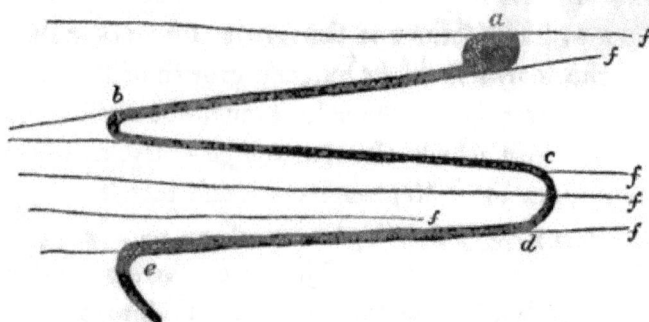

Bird's-eye Plan of the Victoria Falls and River, after a Sketch in the Dublin Exhibition, 1872.
a Victoria Fall ; a b c d e, river ; f f f f f f, joint lines.

may meet a joint-line belonging to another which is better developed and more easily denuded, in consequence of which it will leave its original course and take to another. This is well illustrated at the great Victoria Falls, Africa, where the water flows over into one break, and from that finds its way into a second, a third, and a fourth, so that the river's course below the falls assumes an extremely acute S-shaped form. At home it is illustrated in the ravines occupied by the lower portion of the Suir, the Barrow, and the Nore, in south-east Ireland, where the different streams have excavations along different systems of breaks; while in general the breaks along whose continuations those rivers originally worked are now occupied by tributary streams. A remarkable instance of how a river may leave its original joint-line is seen at the junction of the Suir and the Barrow. The former is the larger stream, yet it has left the course it might have been naturally expected to follow, and now flows in the continuation of the break along which the Barrow came, previous to their junction. Numerous other illustrations could be given; these details, however, are more suited to the body of this book than to its introduction; it is therefore unnecessary to enter further into the subject.

Although valleys seem not to be formed without the assistance of breaks or faults, yet all fault-lines do not form valleys; on the contrary, they may form peaks and ridges. Nearly all the peaked summits among the mountains of West Galway belong to hills formed of quartz schist, and these rocks are traversed by large dykes of hard "fault-rock," a breccia having a matrix of either quartzyte or vein-quartz. As these rocks are harder, and have resisted denudation better than the associated rocks, many of the peaks are on a line of dyke, or at the crossing of two ; while other dykes form the summit of ridges, or stand out as reefs across the hills, or forming shoulders to them.

In other places no traces of the faults can be seen on the surface of the ground. This has been especially remarked during the working of the different coalfields, and it led an eminent geologist to state that he could not see how there should be any connection between faults and surface features, when so many large faults in the coal measures are not discernible on the surface of the ground. The obliteration of such fault-lines may be due to two causes : these lines may be simple " slides," while the associated rocks are so similarly constituted that the denudant (sea or ice) carved them evenly away ; while in other cases the surface features due to faults may have been obliterated by the subsequent deposition of an envelope of drift. That drift often obscures these features is proved by mining and other such workings, as one pit may reach the rock in a few yards, while another in its vicinity will have to be sunk perhaps forty or fifty ; and in a drift country it is generally found that most of the fault-lines, except some that are post-glacial, are obliterated. It might be said that all the post-glacial faults should be apparent; this, however, need not be the case, as the sea subsequently has, in many places, modified the surface of the glacial drift.

In the following pages we will attempt to demonstrate that, in general, valleys are connected with faults or breaks, and that a valley or hollow could seldom have been carved out unless there were cracks, minor joints, or other shrinkage fissures, in which one or other of the different denudants could work. The majority of our examples will be taken from facts observed by ourselves, principally in Ireland, but we will also refer to what other observers have noted in various countries.

PLATE I.

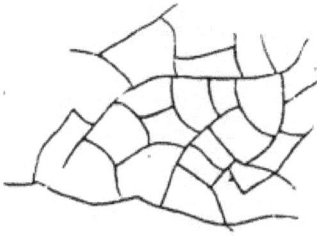

Fig 1

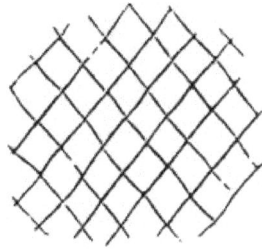

Fig 2

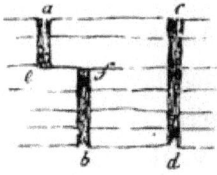

Fig 3

Fig 4

Fig 5

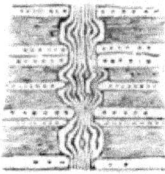

Fig 6

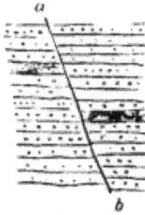

Fig 7

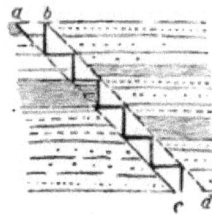

Fig 8

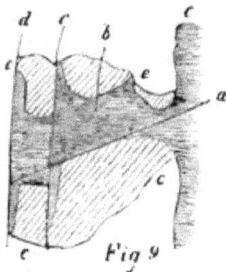

Fig 9

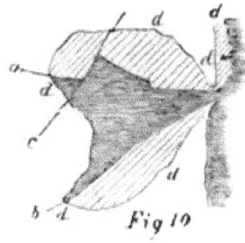

Fig 10

CHAPTER I.

ALL rocks, no matter of what composition, are, in a greater or less degree, traversed by cracks, joints, and faults, and to explain the origin of these is the task that first presents itself. In order to do this, we must either make similar breaks artificially, or watch what is going on at the present day. A recent deposit or accumulation, such as a bank of silt in a river, lake, or estuary, or an alluvial flat unwatered by the lowering of a river, or a drained bog or moor, when exposed to atmospheric influences, will crack as it dries and contracts, sometimes irregularly (fig. 1, Pl. I.), but often more or less regularly (fig. 2, Pl. I.) However, on account of the frail nature of the materials, rain, frost, heat, and the wind may more or less obliterate the fissures, but some will remain open to act as water-carriers if they have a fall; they will thereby be deepened, while the increase in their width will probably be small. Numerous examples of such

proceedings occur, especially in mountainous bogs.
One worthy of record was observed north-west of
Roundstone, County Galway. Here is a flat bog
bounding a lake; and some years ago, during a very
dry summer, a crack opened in the former; this was
taken possession of by the water of the lake, which
deepened and slightly widened it, thus forming a
connection between the upper lake and Lough-na-
Sooderry, so that now the surface of the water in the
former is always four feet lower than it used to be.

Cracks also form in flat bogs, and at different times
they have indirectly caused large tracts of the Irish
bogs to move. One of these movements occurred about
fifty years ago, near Clara, King's County, under the
following circumstances:—The drought of a very
dry summer not only opened fissures in the bog, but
also, as it would appear, in a great measure cracked
it away from the subjacent marl and gravel. Such a
process can be seen on a small scale on any boggy
mountain after a few weeks of very hot sun. The
water seems to have afterwards got under the peat,
and when the inhabitants of the country cut their
turf, they severed the marginal connection between the
peat and the underlying strata, and the whole bog
moved away. The phenomena was thus described by
an eye-witness:—"I remember the bog moving. It
is about forty years ago, the year before the king
[George IV.] came to the Curragh. It was a very

dry year, and the year the king came over was very
wet. I was then a small boy, and lived near the place.
It happened in June, on a Friday during dinner-
hour. The cows saw it first, for they all began
running away from the bog, and we thought it was
the flies were at them; but then the barrows and
slanes began to tumble about, and the bog to move
up and down like the waves of the lake. It moved
as far as the small cabin on the bog that day, but on
the following Monday it again moved across the road
up the side of the gravel ridge (Esker); it tore up the
marl with it, and carried it along. We considered
that it was such a dry year that the lough on the
east side of the bog got under it, down through the
cracks, and floated it away. People from all parts
came to set up tents in the demesne to see it move,
but it never stirred since." In this case it would
appear as if it was the turf-cutters who by their operations had finally given occasion to the bog to move.
But usually in the case of moving bogs, the connection between the peat and the gravel or marl is cut
away by the action of a river or stream flowing at
a lower level, the bed of the bog having a slight
incline towards it; but once the mass begins to move,
its own weight may force a portion of it on to a
much higher level, as in the case of the Esker, near
Clara.

Movements somewhat similar to that now mentioned have taken place on mountain sides when cracks have opened across the slopes, as in these water will collect, till eventually large masses float, and afterwards loosen and slide down into the valleys. A slip of this kind took place on the mountains of Glancastle, near Belmullet, County Mayo, in 1867, and a description of it was written by the Rev. P. Mahon, P.P., of which the following is an epitome :—About forty acres of country, varying from ten to fifteen feet in depth, broke loose from the mountain-side and slid down into the Atlantic, converting what had been a luxuriant dale into a desolate valley. The slip was due to " the great drought of the preceding two months creating a vacancy between the peat and its gravelly substratum. This was filled by the heavy fall of rain on July 19th, by which the peat-covering of an entire mountain-side was raised from its bed, broken up into huge fragments, and caused to slide down the inclined surface, carrying destruction in its course, until it spent its fury in the Atlantic. The slope of the ground was about one foot in fifteen." In this case it was a peaty soil that gave way ; but similar slips may also occur in drift, and they will hereafter be referred to when we are describing bars and lake-basins in mountain valleys.

In alluvium (sand, gravel, and silt) cracks are not so effective in this way as in peats, since the sides shed from the effects of heat and wind or contact with water; yet in some cases what was originally a crack may eventually become a watercourse.

When a section is laid open in a bank by a river, or lake, it is usually found to be stratified, according as materials of different composition, colour, or texture may have been deposited in layers or strata. In such accumulations two distinct kinds of cracks or joints are formed; one set (*minor joints*) are local, and only affect a stratum, or two or more associated strata, while the other (*master joints*) cut through all the layers. The minor joints are, in general, locally systematic—that is, in some layers, only, they will all be regular, and in others all irregular. The master joints on the surface are usually more or less regular, but beneath they may be either regular (*c, d,* fig. 3, Pl. I.), or irregular (*a, c, f, b,* fig. 3, Pl. I.), as they take advantage of the weakest lines in each successive stratum they pass through; or even at times they may for a while run along the stratification, and if there be an open crack at the surface, it may not be continuous downwards, the lower part of the fissure being either to one side or the other of the upper (fig. 3, Pl. I.)

If two master joints cross, they form a greater or less vacancy at the intersection (fig. 4, Pl. I.); but if three or more cross at one point, the opening will necessarily be larger, not only on account of the com-

bination of the fissures, but also because the project-
ing angles of the intervening masses are liable to
break off or crumble away (fig. 5, Pl. I.)   Minor
joints may act similarly, and they are largely instru-
mental in causing reclaimed bog, unless well weighted
with clay or gravel, to return to its original condition. The process is as follows :—After the bog is
laid down in grass, it cracks and opens during hot
weather. The fissures do not afterwards close, but
their sides shed, forming muddy spaces round patches
of grass. Yearly this opening goes on, until eventually the grass land turns into innumerable small
heathery hummocks with soft bog between, moss
and heather having killed out the grass. A similar
process may occur in alluvial flats, except that rushes,
instead of moss and heather, usually succeed the
grass.

In any recent accumulation, one or more layers
may be composed of sand, marl, or some such material
that will "run" when a section is exposed, either
naturally or artificially. Fine sand and the like
usually runs, on account of the water it contains,
moving off and carrying the sand with it. Slips of
drift, due to this cause, may be studied on the coast
of Counties Mayo and Galway, in places where there
are accumulations of boulder-clay drift containing
lenticular layers and subordinate beds of sand. Some
of the latter, when exposed by the sea action, "run ; "

while subsequently cracks form in the superincumbent mass, till eventually, having lost all support, it slides away seaward. On this coast such subsidences are usually not of very great extent, but in a few cases remarked they were considerable.

Streams may act somewhat similarly to the sea; and examples of their undermining and lowering ground may be studied on the south slopes of the mountain group called Slieve Arra, County Tipperary. On these slopes there is a large accumulation of boulder drift, which in places contains subordinate beds of running sand. In the boulder drift the rain and rivers have opened deep ravines, and if these reach down to the sand, it runs, thereby causing large subsidences in the drift. Some of these subsidences are in steps (see plan and section, fig. 18, 19, Pl. IV.), as if the drift when sinking had given way successively along lines of parallel jointing, thereby preventing the sand farthest away from the vent having free egress.

Remarkable instances of subsidence are those due to a bed of sand being carried off by a spring, that may come to the surface, at any distance away from the source, of the sand. As the sand must be subtracted particle by particle, the subsidence usually is gradual, forming more or less bowl-shaped hollows in the surface of the ground. This, however, is not always the case, as the ground may give way

along joints or breaks in the drift, and thus form
hollows with perpendicular sides; but in such cases
the shape of the sides is generally soon modified by
meteoric abrasion.

Something may also be learned from railway and
other embankments, as they often subside on account
of the ground under them being weak.   If the
foundation is bog, the peat will bend down at first
towards the embankment, but afterwards it will
crack in lines rudely parallel, and eventually form
fissures more or less wide, according to the depth the
embankment sinks.   Gravel, in general, when it
gives way, acts very similarly to bog, bending down
to the embankment, and eventually cracking and
forming fissures.   So also do some clays, but others
act like mud, which squeezes out from under the
embankment, and rises in curves.   Clay will stretch
to a considerable extent, and may remain unbroken ;
but mud seems to crack immediately—at first, perhaps, often irregularly, but subsequently, when it has
lost most of its moisture, regular fissures may be
formed in it.   Subsidences may bend the different
strata, but in many cases they occur along a joint,
and thereby break the continuity of the beds or
strata, thus changing a joint into a dislocation or
fault.

CHAPTER II.

WE have endeavoured to point out what can be learned from the study of the formation of cracks or joints in recent deposits; and if from them the student turns to the observation of rock masses, he will find that nearly similar laws appear to have been followed in the jointing of all rocks of derivate origin, no matter to what geological period or age they belong. One marked difference, however, ought to be pointed out ; namely, that in the older rocks, generally the joints are more regular and systematic. There are, however, exceptions to this general rule.

It seems probable that all rocks, when deep-seated, are more or less charged with moisture. This they lose after being subjected to a drying process, whether by atmospheric influences or internal heat. Rocks, when dried, contract in a greater or less degree, according to the composition of the rock; the contraction being rarely equal in the vertical and in the horizontal

B

directions.   Hence in well drawn out building contracts, it is specified that all stones should be laid
on their " quarry bed," as otherwise the settlement
of the masonry will be uneven.   Dry rocks or stones
that have been for some time quarried contract less
than freshly quarried rocks.   It may therefore be
suggested that stones, before being used for building
purposes, ought to be allowed to lose their quarry
water; for if fresh moist stones are put into a building along with well dried ones, no matter how neatly
or finely the joints are laid, these must eventually
open more or less in certain places.   Open joints are
often attributed to bad building, when in reality the
fault may have been in the mixture of stones in different stages of contraction.   This often occurs in
buildings; new fresh quarried and cut stones being
used as quoins and such like, while the intervening
work may be built of old stones, which have been
drying for years in another building.   Different kinds
of stones contract in different degrees.   This subject, however, seems not to have attracted sufficient
attention, as we have been unable to find the records
of any experiments on the contraction of rocks.

Rocks in *situ* must form by their contraction
joint-lines; and as in the recent deposits, so also in
the older rock masses, there are *minor joints* only
affecting one or a few associated beds, and *master
joints* that cut through many strata.   Minor joints

seem to form when rocks are at or near the surface, but some are deep-seated, as joint-lines may form at an old surface, while subsequently newer strata may have been deposited. In the old rocks these joints are usually much more regular than in the recent accumulations; nevertheless, in some beds of limestone a jointing quite as irregular as, or even more so than, that figured (fig. 1, Pl. I.) may occur; and such jointing, combined with subsequent weathering, will change an even surface of limestone into a coarse shingle, or into a number of irregular hummocks; each being surrounded by a deep hollow, and the whole collectively having a somewhat similar appearance to a tussocky bog; both being due to the same causes, namely, irregular jointing and subsequent weathering. Jointing of this kind is not uncommon in places among the limestone crags of Clare and Galway; the interstices between these hummocks varying from a few inches to yards in depth. We may, however, remark that although we have mentioned these localities as places in which these joints may be found, yet they are not characteristic of the limestone, for generally the jointing is extremely regular.[1] A barrel of Roman cement, if left in contact with the atmosphere, will become indurated into stone. Such an artificial rock was used as the corner-stone of a

---

[1] "Mem. Geol. Survey, Ireland," ex. sheets 85, 95, 105, 106, 113, 114, 115, 122, 123, &c.

house; and on the outside it weathered similarly
and had the aspect of an argillous limestone.   On
being broken, there were internally nearly parallel,
more or less regular, horizontal joints, and two sets
of vertical joints, which enable the block to be easily
broken into irregular cubical pieces.   In the cubical
pieces there was a horizontal lining, seemingly due
to the packing of the cement in the barrel.

Joint-lines gradually opening can be studied in an
abandoned limestone quarry, where sheets of rock
have been uncovered, and left subject, for greater or
less periods of time, to atmospheric influences.   Here
it will be found that in the parts longest exposed,
fissures have been produced along the joint-lines;
while in the places more recently stripped, the joints
remain in their original condition.   The fissures are
partly due to meteoric abrasion; but that contraction
has also partly formed them, is evident in the places
where the joints have cut through layers or nodules of
chert, this very hard kind of rock being nearly un-
affected by weathering, yet contraction has separated
one portion from the other.   The connection between
contraction and jointing is very perceptible in such
rocks as hard conglomerates.   In them the surfaces
of division will be found to cut through the
enclosed pebbles, and each part of any pebble will
exactly fit its fellow on the opposite side of the joint,
showing conclusively that the jointing could not be

due to weathering. Nevertheless, the peculiar character of a large proportion of joint-faces seems to indicate that something more than mere cracking consequent on shrinkage has been concerned in the formation of them. This is perhaps particularly well illustrated in granite, and also indeed in the hard conglomerates. The way in which the joint-planes often cut smoothly and evenly through the matrix and the quartz pebbles in such conglomerates, is a perfect mystery, and not to be explained by mere cracking or mere fracture of any kind, as the fracture-surfaces of these rocks is often totally different in character from the ordinary joint-surfaces. Haughton suggests that these surfaces are due to an obscure tendency to cleavage having previously existed in the rocks.

Our colleague, the late F. J. Foot, M.A., paid particular attention to the jointing of the limestones in the barony of Burren, Co. Clare, and of it writes :—

"The main joints, in this district, range between N. 45° W. and N. 45° E., the most prominent being those from N. 5° E. to N. 10° E. These may often be seen running in perfectly straight lines for several miles. Sometimes the piece of rock between two of these is traversed by numerous small parallel joints, extending for the same distance, and cutting up the rock into vertical laminæ. The action of the weather on these will in time wear away this intervening piece, and cause a deep open fissure. Sometimes the main joints assume curved instead of straight courses.

" The cross joints range from E. 45° N. to E. 45° S., the most common being probably about E. 20° S. They are not nearly so well marked as the main joints, never being traceable for any considerable distance, and seldom running even for this short length in straight lines. In many places there are no visible cross joints.

" It is possible that the gentle southerly dip (or, more accurately, S. 5° or 10° W.) is the cause of the north and south main joints (or, more accurately, those bearing N. 5° to 15° E.) being the most prominent. The dip of the surface is the direction in which water will flow; any joint or crack in the rock, therefore, lying in this direction, will be more exposed to the mechanical action of the rain, than one which does not lie in it. In the first case, the water has its maximum velocity; in the second, it will either lie stagnant, or flowing slowly, will exert less mechanical force.

" As we go westward in this district, it would appear that the main joints have more of E. in their bearing. In the island of Innisheer the main joints are most beautifully marked, running across the island, and bearing constantly N. 17° E. There are also numerous main joints bearing W. of N., and cross ones N. of E., but they are never traceable for any great distance.

" The general dip of the surface on Innisheer is S. 15° to 17° W., and the most prominent open joints bear N. 17° E." [1]   This authority ignores the denudation by wind.   This force, however, as hereafter shown (*pp.* 79 *et seq*), must have materially assisted in weathering these joint fissures.

Joints may be very variously developed as to frequency, for at one time the members of a system will be wide apart; while at other times they may be so close as to cut up the rock into slate-like slabs. Cleavage in rock masses is easily determined, but the distinction between minor jointing, when confined to a single bed, and cleavage, is often obscure; in fact, in some cases one phenomenon seems to graduate into the other.[2]

The contraction of rocks is apparent in mineral veins that traverse alternations of soft and hard strata.   If there has been no displacement of the rocks (*slide* or *heave*) in either wall or "cheek," the parts of the same divided bed will be found opposite each other on each side; and if there has been no shift, there can have been no friction or rubbing to

[1] "Mem. Geol. Survey, Ireland," ex. sheets 114, 122, and 123, pp. 21, *et seq.*

[2] W. King, D. Sc. Prof. of Geology, Queen Coll., Galway, has studied the connection between these two phenomena.   To his papers on the subject the readers may be referred (*Report Brit. Assoc.* 1858, p. 83. *Scientific Opinion*, 1870).

wear away the softer beds ; yet the lode will swell in
the soft strata, and contract or " choke " in the hard,
suggesting that during the drying process the argillous beds have contracted laterally more than the
arenaceous (fig. 6, Pl. I.) It may be said that the
rocks may have moved up and down, and finally
stopped in their original position. This undoubtedly
seems to be the case in some instances, but in
most others it is highly improbable. In Salisbury
Crag, Edinburgh, the joints (faults) in the igneous
rocks are mere lines ; while in the associated
derivate rocks, they are marked by a greater or less
thickness of " fault-rock." Similarly in a bank of
silt, we find that the argillous beds contract during
the drying process more than the arenaceous.

Fissures and faults in strata are not solely due to
the unequal contraction of the different beds, as
many of them have been formed during the several
movements in the earth's crust. These latter movements are generally supposed to be due to the interior
portion of the earth cooling and contracting more
rapidly than the outermost shell. The latter being,
in consequence, subjected to powerful horizontal
pressure, which continuing to increase with the pro-
gressive interior contraction, the exterior portion must
eventually yield in places.[1] Other faults may be

[1] " Formation of the Features of the Earth's Surface." By Prof. J. le
Conte. *American Journal of Science and Art.* *Third Series.* Vol. iv.,
pp. 345, *et seq.*

caused by a part of a bed or beds, swelling or expanding, and thereby raising and breaking the superincumbent strata. This latter phenomenon can be studied on a small scale, in the following simple experiment :—Mix some lime into mortar, and after it is made, introduce pieces and fragments of unslacked lime, then spread out the mass, and level it with either a roller or trowel. As the introduced pieces slack, they expand and raise the mortar over them, sometimes forming regular systems of faults (a main fault with transverse branches) ; at other times a dome will first form, but this, if the expansion is great enough, will eventually split, and caught-up portions of the mortar are often elevated, just as we find masses of an older igneous rock raised on a newer. The boundary of these raised portions may be most irregular, yet on examination it seems invariably to be a combination of straight lines.

The widening of master joints, and the consequent formation of fissures, seems sometimes to have gone on gradually, while in many other cases they would appear to have opened at different periods with intervals of rest between each. This seems suggested by finding in many mineral lodes layers in duplicates, each set differing in texture, aspect, and composition, as if each opening had been formed and filled prior to the subsequent expansion. Jukes, however, points out that the layers of different minerals in some lodes may not be due to their having been "successively introduced, since all the substances may have been in solution together, and circumstances having been favourable at one time to the deposition of one substance and to that of another at another time."[1]

But this could not be the case in all lodes; for in some, such as the lode at Maumwee, West Galway, the original lode stuff was metamorphosed contemporaneously with the associated rocks, while subsequent to that action a newer lode formed alongside the older one. Other cases of subsequent contracting are metalliferous strings and veins traversing and cutting through the other minerals in a lode, or a "rib" of ore may run for some distance along one wall and then cross over to the other wall, evidently occupying a crack due to shrinkage.

Mining operations have taught us that there are three distinct kinds of fissures or breaks. These are called by the miners *slides* or *heaves, cross-courses* and *lodes.* A slide or heave is a simple single break in the strata along which the beds on one side have slid down or been heaved up (fig. 7, Pl. I.) A cross-course is a break, the walls or sides of which are separated one from the other, while the intervening space is filled with a dyke or mass of *fault-rock* (fig. 8, Pl. I.), that is, a rock made up of fragments and the detritus of the associated rocks, or of materials

---

[1] "Student's Manual of Geology," a New Edition, A.D. 1862, p. 359.

forced up from below or dropped into the fissure from above; while a lode is a shrinkage fissure, or a fault now filled with minerals and their associates. The fault-rock generally found in cross-courses would appear to be due to the rubbing up and down of one wall of the break against the other, as happens with the sides of a crack formed in a wall during an earthquake; and in general the strata on either side are displaced from their original position. This, however, is not always the case, as in some instances the beds of rock on both sides would appear to have returned to their first positions, cross-courses being known to cross lodes, and yet neither to heave nor slide them. Some of these cross-courses may possibly have been open fissures that were filled from above, but others could scarcely have been so.

Most lodes, originally, were fissures due to the shrinkage of the rocks. These eventually were filled by minerals deposited from solutions; and, as just now stated, it is probable that in many cases the fissures enlarged gradually. At Glandore Mine, Co. Cork, there are three distinct formations occurring together, and forming one great " mineral channel; " namely, manganese-ores, hematite and fault-rock; the latter being the oldest, as the hematite occurs in a lode that crosses and sends veins into the fault-rock, while veins of the manganese-ores are found cutting through both.

That class of granite vein for which Dr Sterry
Hunt has proposed the name of "endogenous veins,"
is evidently allied to mineral lodes, as the rocks
now filling them were deposited from solution.   This
is evident, as most of them are lenticular, that is,
die out in every direction ; consequently they could
not have been filled by matter injected into them. A
character possessed by them, in common with mineral
lodes, is, that in many of them there are successive
layers or ribs along the walls, usually of very similar
minerals, but always differing in texture.   That such
layers, filling shrinkage fissures, have been deposited
successively, can in some instances be proved by veins
and strings branching from the inner layers, not only
into the outer layer, but sometimes even into the
adjoining rocks ; as if the more recent shrinkage had
affected, not only the rock mass, but also those parts
of the vein-rock that had been previously formed. As
slides, heaves, cross-courses, and lodes are closely
allied, one may graduate into another.   The dyke of
fault-rock in a cross-course, or indeed in any fault,
may widen or narrow.   If the rocks, through which
the fault passes, are alternations of hard and soft
strata, there will probably be a dyke of fault-rock in
the latter, while in the former its course may only
be marked by a line or mere parting ; a cross-course
graduating into a slide or heave. This is often
found to be the case when igneous rocks are inter-

stratified with sedimentary rocks; and Henwood, when describing East Cornwall, states that in Mentreniat, "the whole strata seem broken up by a succession of disturbances of a nature between cross-courses and slides."[1]   Somewhat similarly a slide may graduate into a lode, for in alternations of "soft and hard ground," such as shales and grits, the lode will widen in the former, while it may "choke out" entirely in the grits, and only form a simple single break.   Cross-courses and lodes also merge into one another, the fissures having been in part filled with fault-rock, and in part by a mineral vein, or by a combination of both ; the interstices in the fault-rock having been filled with minerals deposited from solution.   Henwood mentions a remarkable example at Mentreniat,[2] and such are not uncommon in other localities.

Most of the shrinkage fissures beneath the surface of the earth seem to be filled with either fault-rock, or crystalline mineral matter, but those formed at the surface are affected more or less by the atmospheric agencies.   If surface fissures occur in level ground, or in ground from which there is only a slight fall, meteoric abrasion will more or less obliterate them; but if they are on a slope, or there is a subterranean drainage from them, the meteoric

---

[1] "Cornwall Geological Transactions," vol. viii., p. 714.
[2] *Ibid.*, p. 713.

debris will probably be removed by rain and runnels as fast as it is formed. Consequently the fissure will not only be kept open, but also enlarged, the denudation generally acting more in depth than width. The consideration of denudation would, however, lead us away from the present portion of our subject; and prior to entering into that, we must describe more fully the connection between master joints and faults.

CHAPTER III.

FAULTS.

It has been pointed out that some faults are due to the unequal shrinkage of strata; others to movements in the earth's crust. The latter, however, are, for the most part, indirectly connected with shrinkage, as it is the contraction of the interior of the earth that causes the movements in the crust. Some faults are due to the subtraction of matter, others to its expansion; as matter may be subtracted either by running water or volcanic action, and expansion may be due to vulcanism or chemical action.

As the composition of the rocks forming the earth's crust varies, not only in passing from one to another, but also in different parts of the same mass (as one part of a bed may be more argillous, arenaceous, or calcareous than another), different beds, and even different parts of the same bed, may contract unevenly. The Caudelaria lode, in the district of Chanareillo, Chili, "is seldom more than six inches wide when it crosses the hornblendic dykes; but though uniting with no other veins, it enlarges at the upper side twelve or fifteen feet, immediately on entering the adjoining limestone,"[1] thereby giving proof that the contraction in the limestone at Caudelaria was much greater than in the hornblende rock. It should, however, be pointed out, that part of the vacancy in the limestone might possibly have been formed by a subterranean stream prior to the deposition of the metalliferous vein, and that consequently the vacancy need not have been wholly due to contraction. If in rock-masses contraction takes place in a vertical direction, there must be unequal subsidences, unless all the rocks are severally homogeneous in the horizontal direction, which is rarely the case. These unequal subsidences must take place along the lines of greater weakness, which will be the master joints; and such movements must break the continuity of the strata, and change the joints into *dislocations* or *faults*.

Henwood's observations remarkably confirm the connection between faults and the uneven shrinkage of rocks. He states, in reference to the recorded examples of lodes in Cornwall and Devon intersected by the same cross vein: "There is no instance in which motion in one direction, and of the same extent, will restore the continuity of every lode so intersected; and such uniform motion will, in fact, produce, in their relations on opposite sides of the cross-vein,

[1] Henwood, *ibid.*, p. 83.

greater discordance than those which at present subsist." [1]

In rocks belonging to the Kainozoic and Mesozoic epochs, the contraction in general has been less than in the rocks belonging to the Palæozoic epoch, while the latter rocks present a marked contrast to the metamorphic rocks. Consequently in the Kainozoic and Mesozoic rocks, faults and fissures may be few and small; in the Palæozoic rocks, they may be numerous and large; while in highly metamorphosed rocks, they are often nearly innumerable.

Faults may be simple or compound. A *simple fault* is that in which the movement has taken place along a single line, and the motion has been so regular that little or none of the adjoining rocks were worn or torn away. In a *compound fault*, there are two or more lines connected with the movement, and a greater or less thickness of "fault-rock" has in most cases been formed. If a movement takes place along a straight or nearly straight master joint (fig. 7, Pl. I.), little or no abrasion can take place; but if a similar movement were to happen in an uneven joint (*b, c,* fig. 8, Pl. I.), the pressure and motion would break off the jutting-out portions of the beds, as represented by the parts included between the dotted lines *a, c,* and *b, d* (fig. 8, Pl. I.) This figure represents a regular joint of this class,

[1] Henwood, *ibid.*, p. 370.

C

while in nature such joints are usually more uneven and often most irregular.

Another kind of compound fault is due to numerous master joints in proximity, running more or less parallel to one another, and all being concerned more or less in forming the faulted ground; or the joints may branch off from one another, and afterwards join again into one, as represented in the accompanying section and plan (figs. 20, 21, Pl. IV.) The displacements may be tolerably equal along each joint-line, thus forming what has been called a *step-fault* (fig. 21). The movements, however, are often most irregular, especially among the metamorphic rocks. Long strips of country are sometimes met with, occupied by "fault-rock," or strata so jumbled up and mixed together, that it is nearly impossible to separate one kind from another; or, as is not uncommon, the hard varieties of rock have crushed up those that were softer, and masses of the first will be enveloped in the debris of the others.

Such a tract of faulted country was observed in West Galway, extending from Dogs' Bay on the south, northward past Clifden to Cleggan Bay. This broken ground is sometimes more than a quarter of a mile wide, while in places it is represented by a dyke of fault-rock only a few yards in thickness. This is altogether due to the disposition of the lines along which the different movements have

taken place—they in some places running rudely
parallel to one another, in others branching off; or
all the movements may be confined to one narrow
line. That movements or faults are prone to occur
over and over again in nearly similar lines seems
proved here; as in this tract between Cleggan and
Dogs' Bays there are proofs of two; while probably
three or four movements have taken place. The
first, post-silurian, but prior to the rocks having been
metamorphosed, as the fault-rock, after it was formed,
was altered along with the associated rocks. The
second later, but still pre-carboniferous. Of this movement there are here no positive proofs; as there are
no carboniferous rocks in the immediate vicinity; but
movements during this period are known to have
taken place in other parts of the district. The *third*,
post-carboniferous; the exact time of this movement
is uncertain. The *fourth*, post-glacial.

As in master-joints, so also on fault-lines, open
fissures may form, when deep-seated, to be filled with
minerals, but when at the surface, to remain open
until they are more or less modified by meteoric
abrasion. As in the recent accumulations, so also
among the older rocks, a vacancy of greater or less
magnitude will probably be produced by this agency
at the juncture of two or more fissures (figs. 4 and
5, Pl. I.) The open space due to contraction may
not always be continuous in one line, especially in

contorted strata; but it may be shifted to one side or
the other, if the line of least resistance in part coincides with one of the lines of stratification (fig. 3,
Pl. I. *a*, *e*, *f*, *b*). Open fissures forming caves, due
to such a cause, are not uncommon in some localities. They were noted in the sea-cliffs of Kerry and
Cork, while Henwood, in his description of the mines
of Chili, figures lodes that occur in fissures of this
class.[1]

Landslips in recent accumulations have been
described, and very similar slips may occur in the
older rocks when at the surface. Some of these are
due to subordinate beds "running;" others would
appear to be caused by the subsidence of a tract of
country along a line of master joint or into a subterranean vacancy, the origin of which we are unac-
quainted with, and can only guess at. An example
due to the running of a bed is the well-known land-
slip in the neighbourhood of Axmouth, Dorsetshire.
Here it is quite apparent that the rain water descending through the joints in the chalk and green sand
softened the underlying lias clay, which "ran," and
this, combined with a slight dip of the strata seaward,
induced a large tract of country about a mile long to
slip. From an examination of the land margin of the
"slip," it is apparent that the first and all subsequent slips have taken place along joints more or

[1] *Ibid.*, p. 82.

less parallel to the coast-line, which broke the
connection between the seaward portion and the
mainland; consequently, as fast as the foundation
gave way, the slips followed.

In Antrim, Ireland, along the chalk escarpment,
very similar slips have sometimes occurred. Open
joints have, at first, gradually formed in the chalk,
through which moisture has percolated to the under
beds, softening and carrying them away, till eventually
large masses of rock slide out into the lower country.
In the whole of Ulster it is hard to determine
whether the great slips, the remains of which are
apparent, are solely due to meteoric action, or
whether they took place when the sea washed the
base of the escarpment. During the present time,
and for ages past, such slips do not appear to have
taken place; but it is quite evident that if the great
slips took place when the sea occupied all the low
country, other minor slips must have taken place
subsequently, probably immediately after the elevation of the land.

Along these escarpments two principal systems of
joint-lines may be observed, one rudely parallel and
the other nearly perpendicular to the line of escarpment; and to these, in a great measure, is due the
different features the basset of the chalk presents. If
joint-lines are nearly parallel to a line of escarpment,
as they open the water lodges in them, and even-

tually prizes out masses of rock. In our special case,
however, it also acts on the soft lower beds (sand and
marl), and either carries them away, as may be seen
in the springs flowing out of the sand beds, or makes
them so soft and slippery that the superincumbent
mass has no foundation, and its connection behind
being already cut off by the joints, it must slide out-
ward.   If joint-lines are perpendicular to an escarpment, the action is quite different, as all drainage
takes place along these lines, and excavates ravines
with abrupt hills between ; and if these joints are near
together, the results of weathering in chalk will be
the carving out of " pillars," " needles," and such
like. Such a weathering does not, however, prove that
the escarpment is due to meteoric abrasion, as marine
action may work very similarly on joint-lines perpendicular to it.

Of landslips in Mount Sirban, India, Wynne thus
writes :—"The Nummulitic formation, as usual among
the border hills of the Punjâb, is chiefly composed of
massive gray and blackish limestone, here alternating
with thick zones of dark-coloured shale. Its thickness is very great, its stratification violently contorted,
and it possesses the same features, commonly ob-
servable, of profound gorges and ravines excavated
in it, high cliffs formed, and, upon slopes, of great
masses having subsided so as to produce complications

of slippage often exceeding in amount the throw of genuine faults."

Some slips are evidently due to a sliding along an oblique master joint. Such a slide has taken place on the north side of Bantry Bay, County Cork, immediately west of Pulleen Harbour, a tract of country a mile long, and about a furlong wide, having slipped out seaward. Artificial landslips are caused by coal mining and the like; large areas of land subsiding into the vacancies left by the subtraction of coal, iron ores, clay and other minerals; while natural subsidences have taken place during earthquakes, such as that which took place at Lisbon.

Landslips teach us how faults on a small scale can be formed at the surface of the earth, and from them we can reason on the formation of most of the deep-seated faults; all being due to movements along lines of least resistance, which must be master or other joint-lines.

CHAPTER IV.

THERE are various agents to whose operations, combined with the effects of contraction or shrinkage (joints, cracks, faults, and the like), must be attributed the diversified conformations of the earth's surface. These carvers of the earth are—*the sea, ice, the sun, cold, heat, wind, rain, rivers,* and *chemical action. Cold* in this classification includes ordinary frost; while *ice* refers only to the work done by glaciers and icebergs. *Heat* and the *sun* also form distinct headings. These two agents are kept separate, because their modes of action are entirely different ; for although the latter is a source of heat capable of forming surface cracks and other fissures, yet many of the deep-seated joints must be due to the effects of the internal heat of the earth. None of these workers appear capable of doing much work when unaided. The sea cannot act effectively without the sun, chemical action, and wind, to open the cracks, joints, and the like, and to raise the storm and waves. Ice can do little work unless the rocks have been previously affected by joints or other divisions. The sun acts in combination with all the other denudants; and in a greater or less degree, so do cold, heat, wind, rain, rivers, and chemical action. The operations of the seven last-named agents are so intimately connected, that in general they are classed together under the names of *subærial* or *atmospheric denudation* and *meteoric abrasion;* but the individual work of each is peculiar, and a short description of them will be given hereafter. Although the names subærial and atmospheric denudation are very generally used, yet neither seems unobjectionable ; as, strictly speaking, *all denudation is subærial,* and *all is more or less directly due to atmospheric influences.* Difference of temperature causes the wind, the oceanic currents, and the glaciers, also the evaporation that eventually forms the rain and rivers.

It should be borne in mind that ice and the sea as agents have this relationship: they are, for the most part, *mechanical workers dislodging masses of rock,* which are by the former ground up against one another, or against the sides and bottoms of the valleys occupied by glaciers ; and by the latter broken, by being rolled over one another. Meteoric abrasion, on the other hand, ordinarily only *disintegrates rock surfaces,* preparing matter to be carried away by the sea, ice, wind, or rain and rivers.

Some authors delight in describing the sea as throwing its breakers loaded with stones and shingle against cliffs, and thus wearing down even the hardest rocks. The idea of "nature's artillery" battering down a cliff may be very poetical, but it involves a misconception as to fact. The whole of the west coast of Ireland is open to the full force of the Atlantic waves, which often rise hundreds of feet; "blue water" having on one occasion carried away the water tanks at the upper lighthouse on the Great Skellig, which, according to the ordnance map, is 380 feet above mean tide level;[1] while spray has been driven clean over the island of Valencia, so as to wet the windows in Knightstown; the cliff at the west of the island being from 500 to 700 feet high; yet in no place on the west Irish coast did we remark this "battering ram" process going on. On the contrary, in those places most exposed to the waves, the seaweeds usually grow luxuriantly, clothing these rocks with a mantle. Abrasion by the sea will take place in cooses, guts, and the like, but this is due principally to the "back-wash." Stones when carried in by the waves fall on a cushion of water, and the force of the fall is broken ; but as the waves retreat, they roll in the back-wash, over and against

[1] It is remarkable what a height "blue water" will rise on sea rocks, and against cliffs. A wave of the height here quoted, and of sufficient magnitude, if it came in on most parts of the coast of Ireland, would be higher than three quarters of the island.

one another and the rocks.   Consequently it is only in cooses, or such like confined places, they are capable of acting as abraders.

If the sea be watched during action, its principal work, even in the softest shales, will appear to be due to wedges of either air or water forced into cracks, crevices, and other vacancies, by which rock masses are prized out, and high cliffs are thus undermined and brought down.   The sea waves seem never to work but in conjunction with meteoric abrasion, which is often more rapid than the marine worker.   Consequently many cliffs are not perpendicular, meteoric abrasion wearing back the uppermost portion more quickly than the sea can quarry away the base.   Under favourable circumstances the sea may form perpendicular cliffs, such as those on the coast of Clare, where the sea readily carries away the thickly jointed shales and grits of the coal measures; while on account of the horizontality of the bedding, the rocks can stand with a vertical face, as if they were blocks of masonry in a wall.   On the west coast of the Arran Isles, Galway Bay, the sea has formed peculiar cliffs, and as these have been examined in detail, they may be described.

From the N.E. shores of the Arran Islands the land rises in a series of eight cliffs, or huge steps, which form continuous terraces, while from the summit of the island there is a gradual fall south-

westward, ending, however, at the sea-board in cliffs
which, at the present day, are being formed by the
Atlantic Ocean.

The previously mentioned steps, forming continuous
terraces, are several miles in length; two of them
extending from the south-east to the north-west of the
great island. These, the third and fourth, reckoning
from the highest terrace, can be traced from Benaite
at Gregory's Sound, along the flanks of the valleys,
until they join into the sea-cliff, about half a mile
from the north-west point of the island.

The sea-cliffs on the north-east side of the island
are low, and are often replaced by strands, or shingle-
beaches. On the south-west they have taken a
definite character, being usually perpendicular, and
often over fifty feet in height. However, at the
north-west point of the island, under the shelter of
the Brannock Islands, there is a heavy shingle-
beach. From the north-west point, south-eastward
to Gregory's Sound, the cliffs are either perpendicular
or terraced. From Mweeleenareeava, a little south
of the Brannock Islands, to Doocaher, except for a
short distance at the "Blind Sound," the cliffs are
perpendicular, although at the base of some of them,
as will be hereafter mentioned, there are sea-terraces
or steps below the high water mark of spring tides.
At Doocaher the cliffs are about 100 feet high, and
from that towards the north-west they gradually rise

to 234 feet at Corker; from which they lower by
degrees to the "Blind Sound;" but north-west of
this, at Dun Aengus, there is an ordnance height of
265 feet, and they attain their greatest altitude (300
feet) about a mile farther north-west, a little south-
east of Polladoo. From Doocaher towards the south-
east, to the point called Illaunanaur, there are sea-
terraced cliffs (excepting a few short breaks), which
are surmounted by a rampart formed of large blocks.
This latter is called by Professor King, Galway,
" The Block Beach."

The cliffs on the west of this island are peculiar,
as in places they are terraced by the waves of the
Atlantic. Moreover, some of them are surmounted
by the " block beach." This peculiar accumulation
of blocks does not occur at all on the north-east
shore, while to the north-west it was only found at
the point due east of the north point of the Brannock
Islands. On Inish-Eeragh, the westernmost of the
Brannock Islands, there is also a block beach which
is thus described by Captain Bedford :—" On all but
the eastern side there is a margin of massive blocks
of limestone, upheaved by the violence of the sea, and
which now form a sort of barrier against its farther
encroachment. . . . . . The highest part of
the island is the summit of the upheaved beach
at the north-west side, which is 36 feet above the
mean level of the sea."

The nörth part of the north-west coast of Inish-more, as before mentioned, is a perpendicular cliff which either rises sheer up from the sea level, or has at its base a few steps.

The vertical .cliff seems to be caused, in a great measure, by vertical *master joints*, some of which cut through all the visible beds, while others only reach the shale beds, that are found in these limestones at the base of some of the inland terraces or steps, and also as subordinate beds in the limestones that form the south-west sea-cliffs.   In the former case the cliffs are perpendicular down to the sea, while in the latter there are steps or sea-terraces at the base.

Beginning at the N.W. to examine these cliffs, the observer will find that immediately E. of Mweeleenareeava there are master joints ranging N. 70 W., which extend down to the shale beds which inland lie under terrace No. 5; and as the shales at this place meet the sea at about half neap tide, there is a perpendicular cliff above this level, while below it there are steps.

From this for nearly a mile towards the S.E., the master joints bear E. and W., or a few points on either side of that line, and as they extend below low water level of neap tides the cliffs are perpen-dicular.   S.E. of Polladoo there are four sets of steps at the base of the cliff, and the note made on the ground is as follows:—"Cliff over 250 feet high. Two shale beds; the cliff rises perpendicular from the highest. These shale beds are supposed to be the continuation of these under terraces, Nos. 2 and 3." South of Portmurvy there are from four to six of these sea-terraces, and the cliff is less than 50 feet high. South of Gurtnagapple the cliffs are low but perpendicular; hereabouts, nearly E. and W. master joints occur about two yards apart; and as the sea undermines the cliff, masses of rock, tons in weight, which are disconnected by these joints, topple over and fall, forming a breakwater at the base of the cliff. This breakwater extends for about half a mile.

At Corker there is a perpendicular cliff formed by E. and W. master joints. In the face of the cliff there are two shale partings about 40 feet asunder, the lower being about 60 feet above half neap tide. They seem to be the representatives of the shales under terraces Nos. 3 and 4.

The small promontory called Nalhea is bounded by N. 15 E. master joints; here the shale beds have dipped below the sea level.

South-east of Nalhea there are four or five sea-terraces at the base of the cliff, and at Whirpeas the cliff is about 140 feet higher than the level of neap tide, a shale bed occurring about 40 feet above that level. To the east of this, at Pollnabriskenagh, the limestones are traversed by E. and W. master joints,

and the sea yearly causes great destruction of the
rock.   This cliff, which is about 100 feet high, is
undercut at the base.   At Bensheefrontee, the point
a little N.W. of Doocaher, the "block beach" sets
in and extends to the S.E. point of the island, having
only five small breaks in it; three at the cooses or
small bays in the vicinity of Doocaher, one at the
coose called Doughatna, and one about 40 yards wide
at the Glassen rock—in all of which places the base
of the cliff is undercut, while that part which is surmounted by the "block beach" is stepped.   However,
although it is undercut and forms a cave at Doughatna,
yet below the cave there are six very low steps.   The
highest part of the cliff on which this beach occurs
is in the vicinity of Doocaher, and about 100 feet
above the sea level; while the lowest part, a quarter
of a mile west of the Glassen rock, is about 35 feet.
The S.E. point of the island, where the beach ends,
may be 100 feet high.

   These steps at the base of the cliffs are usually from
four to seven in number, seemingly having been cut,
one by low water of spring tides, another by low
water of neap tides, another by high water of neap
tides, another by high water of spring tides, with one
or two intermediate steps when the limestones are
thin bedded, and the systems of joints do not extend
through more than one or two beds.   However, in
places they are modified by master joints, along which

cooses are cut; or perhaps two or more steps will merge into one, according to the number of beds through which the joint penetrates. At one place, east of Carrickurra, there is a step, above high water of spring tides, on which the " block beach " rests; at this place the cliff is about 50 feet high.

The stones forming the " block beach " are cast up during the winter gales, and some of them are of a considerable size. A little south of Doughatna the following observation was made :—" Great quarrying seems to be going on here during the gales. Blocks 30 × 15 × 4 feet tossed and tumbled about." And again, half way between Doughatna and the Glassen rock there is this note :—" A block 15 × 12 × 4 feet seems to have been moved 20 yards, and left on a step 10 feet higher than its original site." East and west of the Glassen rock there are two caves which run for a considerable distance inland, and connected with both are " puffing holes."[1] The western puffing hole is 85 yards from the sea margin, and the eastern 33 yards. On the north-east side of the latter there is a small " block beach," the blocks in which have all the appearance of being yearly tossed about by the waves, while more are added to it, and we may suppose some sucked into the abyss below.

[1] A perpendicular hole connected with a horizontal cave into which the sea has access. During high tides and heavy gales, the compressed air in the caves drives up the water in great puffs of spray through the perpendicular hole.

D

On the west coast of Inishmaan, or the middle island of Arran, the sea-cliffs rise in steps, as they are followed southward from Trawtagh, until at Aillinera they reach their maximum height (nearly 200 feet) ; southward of this they gradually fall to nearly the sea level at Allyhaloo, the S.W. point of the island. Immediately south of Aillinera, at a height of about 170 feet, there is a "block beach," which is continuous from this point round the west and south-east sides of the island.

South of Aillinera the "block beach" is on higher ground than that on which it now occurs on any of the cliffs of Inishmore. The cause for this is not apparent, unless it may be that hereabouts the tides run against the cliffs with greater force than elsewhere, and thereby the waves may be enabled to hurl the blocks to a greater height.[1]

The sea would appear to act more vigorously on a coast-line that is sinking than on any other; but, in such a case, it should be remembered that the cliffs or rocks, prior to having been exposed to marine action, had for years, perhaps ages, been subjected to atmospheric influences that loosened and disintegrated them, leaving them ready to be easily quarried and carried away.

The rocks above low water mark, on a coast-line, are daily exposed to atmospheric denudation, and its

[1] "Mem. Geol. Survey, Ireland," ex. sheets 104 and 113.

influence on soft strata is very apparent; while in places it seems to be the principal denudant, the sea acting chiefly as a carrier.

Margining the sea in south-east Ireland, there extend, for miles, cliffs of marl which are being constantly worn away, yet the sea seems to have very little effect on them. Heat opens joints in the marl, thereby leaving it a prey to meteoric abrasion; while the principal work the sea does is to act as a carrier, and remove the debris brought down within its reach. As in the case of the formation of chalk escarpments previously mentioned, so here also, there are forming two systems of joints that regulate the denudation, one being nearly perpendicular, and the other nearly parallel to the coast-line. In the first, meteoric abrasion widens the joints and denudes the marl into transverse ridges or pinnacles, the joints being near together; but if the joint lines are nearly parallel to the coast-line, they become frequently filled by the rain, and masses of the marl are forced to slide outwards. Thus the denudation of the cliff takes place from the top downward, and not from the bottom upwards, as ought to be the case if the sea was the principal worker. Extreme *marinists* would claim these cliffs as seawork, and if eventually the land is elevated and they form an inland escarpment, extreme *subærialists* would point to them as due to meteoric abrasion; while neither force unaided

could have made them. Sea work alone would make
but little impression on the marl, and the atmospheric agents, unless a river flowed all along at the
base of the cliff, would change them into a slope ; but
when both work hand in hand, meteoric abrasion
prepares the marl to be easily carried away by the
sea. If such cliffs are only examined during wet
weather the connection between the cracks and the
denudation is very obscure, as the rain and moisture
obliterates the cracks. Such places should first be
examined during fine dry weather, when the different
systems of cracks can be plainly seen and studied;
and if subsequently the same place is explored in
rain the mode of work of the atmospheric agents is
apparent.

Very similar slips to those just described, the
result of a similar cause, take place in railway and
other artificial cuttings through marl, clay, and the
like. These, from what we learn in the cliffs, might,
in a great measure, be prevented by making artificial
joint lines (cuts) in them, perpendicular to the line
of railway, which would act as drains to let off all
the water from the transverse joints, and thus
remove the cause of danger.

In some climates, hardened gravel and sand, sandy
clay, or any other clayey drift, may form perpendicular
cliffs to a sea margin ; the sea action working away
the lower portion as fast as meteoric abrasion denudes

back the upper. In such accumulations the assistance
rendered by cracks and other shrinkage fissures seems
to be at a minimum, as marine action at the base of a
drift cliff washes out the clayey matrix and carries
away the rest, while the upper portion is disintegrated,
causing the contained pebbles and blocks to drop out.
If the cliff is exposed to the sun or a dry wind, the
surface of the drift will necessarily crack; but regular
systems of joints are rarely formed. In some places,
however, generally with wide intervals between each,
there are joints or faults. These, if perpendicular, or
nearly so, to the line of cliff, are being worked along by
the sea and the atmospheric agencies, and form cooses
or small bays. Soft sandstones, conglomerates, or the
like, may be worked similarly to hardened gravel and
be cut into perpendicular cliffs.

In the vicinity of the sea, meteoric abrasion would
seem to be more active than inland. On the coast
of Cork the slate beds at the coast are much more
weathered than elsewhere, since the sheet of ice that
once covered that county has melted away.   Inland
the slates are ice-dressed, but in sea-cliffs the rock is
fast being denuded along the joints and cleavage lines,
so that the surfaces of the rocks may be compared to a
number of knife edges ranged alongside one another.
On the Arran Islands the limestone has weathered
away from four to six inches since the glacial period,
as proved by the unweathered pedestals of limestone

under the erratic blocks, while inland similar pedestals are seldom three inches in height.   In Galway it is known, by the veins of segregation (granityte) in the granite, that rocks of this class have weathered in the proportion of two to 'one, as they are respectively at, or away from, the sea.

The sea also appears to affect the colour of some rocks. In Kerry, inland, the purple rocks weather red, and those of a green colour, yellow. At the coast the first become a whitish blue, while the green only seem to be given a brighter tinge; and on the north of Galway Bay the Felstones near the coast were remarked to weather into a colour different from that of the same rock when inland.   Marine action is often represented as forming a plain of denudation ; but marine plains are not necessarily formed by denudation, as they seem to be as often *due to filling up* as to denudation.   This is evident when tracing any old sea margin, such as the 350 feet sea-beach in Ireland, which is more often marked by the terraces due to filling up than by the steps cut in the hillsides.   To form a true plain of denudation the land ought to be perfectly stationary ; and then, the active part of the sea being at the level of the breakers, a plain may be cut ; but even then there are so many different causes to affect it, that rarely off a coast-line will a true plain of denudation exist.   If the sea is working on homogeneous or nearly homogeneous rocks, especially if the bedding is nearly horizontal, it can cut a plain; but if the rocks are of different hardnesses, and are dipping at high angles, no such work can be accomplished. In the first case the coast-line ought to be in long gradual curves, but in the other it must be quite irregular; numerous headlands and bays alternating; the headlands indented by cooses and guts; while islands, carricks, carrigeens, and skelligs lie off the coast and dot the bays, and the soundings are quite irregular—in some places being deep, in others shallow.

The deeps and shallows off a coast-line seem, in a great measure, to depend on the dip of the rocks, and appear to follow recognisable laws: rocks dipping seaward at a low angle form low shore-lines and shallow soundings; rocks dipping inland form perpendicular or overhanging cliffs and deep soundings; while highly inclined or perpendicular strata striking out to sea form most irregular coast-lines, consisting of bays and headlands, with numerous islands, rocks, and shallows lying off the coast, according to the hard or soft nature of the different beds of rock. These rules may be altogether modified by either jointing or cleavage; as one or both of these structures may be more conspicuous than the stratification, and direct the operations of the denudant.

Escarpments in general follow the basset or outcrop of beds, and the base of an escarpment is often inclined in the direction of its length. From this the subaerialists argue that escarpments must be formed by meteoric abrasion. This, however, does not appear to be sound reasoning, as the same cause which might induce meteoric abrasion to work along a line of beds, would also similarly direct marine denudation. On any coast-line, if the strata are more or less flat, the sea works principally along one bed, following it up and down; and even if there are anticlinal curves, it will follow the beds over them, forming a point or headland; and in such places the sea seems to act more vigorously; as the waves dash up the slopes with fury, carrying blocks of stone, and the latter in the back-wash abrade the rocks, thus aiding the mechanical force of the sea. Or the sea action will similarly follow up a bed that is striking away from the coast-line, and rising out above the normal level of the waves; in this way escarpments will be cut by the sea at higher levels than the true coast-line. Farther inland, the meteoric abrasion may also be acting along these beds, and forming an escarpment, so that, if the land rose, there would be one continuous scarp, in part due to the sea, in part to atmospheric influences. This is well exemplified in the previously described cliffs of Arran, Galway Bay; as there the sea is forming cliffs based on certain shale beds, while, in the interior of the island, meteoric abrasion is acting very similarly on the same beds. It may also be seen in numerous places on the coast of England. The sea, nearly everywhere, tries to follow the softer beds under the chalk, and would do so but that the inhabitants have for ages placed barriers to protect these soft parts of the coast from the ravages of the sea, but the chalk they can leave unprotected; while inland, meteoric abrasion has modified the chalk cliffs into slopes, and is said to be also wearing away the underlying softer beds. In Ulster, Ireland, there are also chalk escarpments, due both to marine and atmospheric denudation. This subject will, however, have to be again referred to when we are comparing submerged and elevated sea-formed valleys.

The rock, or rather the debris of the rocks (*Fault-rock*), included between the walls of a fissure, is in many cases softer and more easily denuded than the associated rocks. The waves will work with vastly greater effect in such fissures than on the adjoining rocks, and thereby excavate long narrow guts that may be of great length, although only a few yards wide. Many straits have evidently been so formed, the sea working along, and excavating out a dyke of fault-rock across a promontory, thereby eventually forming an open fissure, and dividing one portion of the land from another. Some of these guts are open to the sky, but others may occur as long,

narrow caves; the latter instances proving that the
sea is capable of working along the fault-rock much
faster than meteoric abrasion.[1]   Some of the deep
narrow straits in South-West Cork are remarkable, as
the old walls of the fissures still exist in part, showing the original hade or underlie of the fissures.
This is well exemplified in the Crowhead promontory,
barony of Bear; the shore of the mainland and the
cliff on Crow Island, hading S.W. at an angle of 75°
(fig. 13, Pl. II.), as if the two had been separated by
a clean cut.

In conclusion, there is one effect of the sea which
may be mentioned, though but indirectly related to
the present subject—namely, its power of producing
hot and cold currents of air, as these materially affect
the land on which they blow.

---

[1] In places on the west coast of Ireland, may be found two caves,
one above the other, the lowest being now in process of excavation, while
the higher one was cut when the land stood at a lower level.   Would
not such double caves suggest the idea that the land must have
risen quickly between the formation of the two? for if the rise was
gradual, the intervening rock should have been excavated away.

PLATE II.

Fig. 11

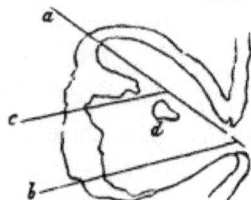

Fig. 12

Fig. 13

Fig. 14

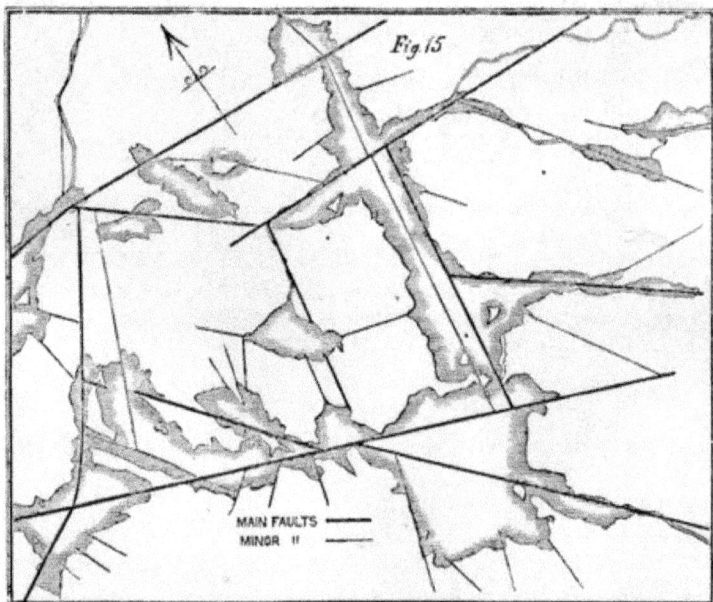

Fig. 15

MAIN FAULTS
MINOR "

MAP OF LOUGH CONGA.

CHAPTER V.

GLACIAL denudation, seemingly, ought to be considered a different action to meteoric abrasion, as these two forces work so differently. A glacier has been said to belong to the atmospheric denudants, as it is a river of ice, formed from snow; but compare the work a glacier and its tributaries can do with that of a river and its tributaries. The tributaries of the first are beds of snow and ice, while those of the latter are all the chemico-fluvial forces combined. The glacier and its tributaries work altogether by mechanical action, being confined to planing and triturating rock, and quarrying, by tearing off masses of rock; while the chemico-fluvial denudation is due to chemical and mechanical action combined. The latter is quite different to that exercised by ice, in this respect, that it is principally carrying away materials loosened by chemical action, the only trituration it accomplishes being that done in river-rapids and mountain torrents; but it does not appear capable of planing down a rock; and ice, as icebergs, may possibly be able to denude the sea-bottom, on which none of the other forces can act, except, perhaps, very strong sea currents. Many rocks, such as doleryte, dioryte, diabase, granite, some limestones, &c., are physically hard, while they are chemically soft; others, such as slates, schists, clays, &c., are physically soft, while they are very little susceptible of chemical decomposition. The former, therefore, are easily denuded by meteoric abrasion, and the latter resist it. Also while a planing action, such as that exercised by ice, would be resisted by physically hard rocks, it could easily excavate and cut away the soft ones; and are not these the results found in nature? In a country dressed, planed, and etched by ice, all the physically hard rocks have formed features from a few feet in height to hundreds or more; but since the ice has disappeared the meteoric agencies have had their sway, and these rocks are now weathering fast; while the chemically hard but physically soft rocks—those that under the old régime suffered most—are now almost perfectly unmolested.

Ice action, similarly to marine action, as before intimated, is principally a mechanical worker, wedging out and quarrying masses of rock. It, however, seems to have a greater power than the sea in one respect —viz., that it is capable of lifting up blocks out of their places and eventually carrying them away. In West Galway, during winter, the water will freeze under large blocks of stone and lift them, though tons in weight; now in a glacier the lifting power ought to be much greater, when the ice is not only under the blocks but all round them. Fractured rock, such as that which fills a fissure, is more liable to be denuded and carried away than unbroken rock; therefore, if ice moves over ground traversed by fissures, especially if the direction of its movement coincides with the bearing of these breaks, the ice, as it moves along, must catch up and carry away all the loose portions; and if parts of a dyke of fault-rock are softer or looser than the rest, the ice will scoop out hollows in those places. If two or more fissures cross one another at one or different points, the rocks in the vicinity of the points of contact will be broken, and at such places ice should be able to work with facility and excavate out hollows, the size of these, both as to depth and width, being in accordance with the amount of rock ruptured and the magnitude of the moving glacier that acted on them. From this it is apparent, that when fissures cross one another there would be a large irregular basin opened out, while if two fissures with broken rock between ran nearly parallel, the excavation would be long and narrow.

The work done by glaciers on their beds may appear at first sight considerable, as the rivers always flowing from them are ever turbid and white with the silt, formed by the trituration of the blocks and fragments of rock in the ice grinding against one another, and the sides and bottoms of the valleys.[1] Moreover, this work is never ending from year's end to year's end, as long as the glacier lasts. On consideration, however, it is apparent that the solid matter contained in glacial rivers, is probably far from giving a true indication of the real work done by a glacier in abrading the rocks it passes over; as the silt held suspended in the waters of the glacial river is not solely due to the abrasion of the rocks passed over, since much, if not most of it, may have been formed from the blocks in the glacier abrading one another. Some of these blocks may have been lifted by the ice and carried away by the glacier, but many of them may have fallen from the adjoining hills on to the glacier, having been dislodged by meteoric action, to find their way afterwards into the body of the glacier through *crevasses*. Also innumerable particles of rock are carried on to the surface of a

---

[1] Hayes says, "The ice was perfectly pure and transparent" (of "Tyndall glacier"), "and yet out of its very heart was pouring the muddy stream of which I have made mention."—"The Open Polar Sea," p. 436.

glacier by wind, and all such auxiliaries to quantity must be subtracted from the real amount of actual work done by a glacier on its bed.

Ice work is somewhat similar to sea work, in that both agents are principally mechanical workers; yet marine action appears incapable of performing as much denudation on ice-dressed rocks as elsewhere. This is very apparent west of Spiddal, Co. Galway, and in other places on the coast of Ireland, where ice-dressed rocks have resisted the action of the sea, although exposed thereto for ages. This would appear to be due to the ice having come after meteoric abrasion, and removed all the rocks and debris that the latter force had loosened, and left ready to be quarried and carried away; also to its having ground the remainder into smoothly-rounded forms, over which the waves would sweep with little effect, there being few roughnesses or abrupt inequalities to offer resistance, and to afford them a handle by which to work. Subsequently, but prior to the sea being capable of executing any great quantity of work, the rocks would again be exposed to atmospheric influences to open up the cracks and joints. The action of ice would, after a time, even render the rock less susceptible to itself, for when the loosened portions of the rock are gone, the pressure and friction of the glacier would be calculated not only to smooth and polish it, but also to harden what remained, as happens in polishing a stone. At first the polishing putty removes all loose particles, but afterwards it not only smooths but also hardens the surface, and we find that polished marble bears the weather longer than a plainly cut limestone, or a sawed sandstone better than a chisel-dressed one.[1] The latter case is well exemplified in buildings, such as those of the colleges in Oxford. There it may be observed that the sawing of the sandstones has formed on the outside of the blocks a thin sheet or coat, impervious to the weather; consequently the weathering does not take place from the outside surface, but has to work sideways in under this shell of rearranged particles; but once they peel off, meteoric abrasion gradually wears away the rest of the stone. The action of ice, however, would seem to be more like that of polishing, as the hardening process appears to extend much deeper into the rocks, and instead of forming only a hardened shell or envelope, the effect seems gradually to die out in depth. If, therefore, a glacier is formed in a valley, the rocks of which have been for years exposed to meteoric abrasion, the work done

---

[1] In burnished metal there is undoubtedly a rearrangement of particles, and this seems to be so in highly-polished marble also, but it has been questioned if such can be in ice-smoothed surfaces. We have, however, carefully examined into the subject, and find that in sandstones, such as the carboniferous sandstone of Mayo, a thin shell of rearranged particles forms on the surface of the rock. In such rocks as limestone there is a rearrangement and hardening gradually dying out downwards, while in such rocks as granite we could detect none.

at first must be large ; but after the disintegrated
and loosened rocks are removed, the effects of the
ice must gradually lessen, as its own polishing of
the rocks it passes over destroys its opportunity of
exercising its abrading powers. That this is the
case seems proved by two, and in places, even three
or four, distinct systems of striæ being preserved on
one and the same rock-surface; these having been
cut evidently by ice-streams which flowed at different times in different directions, and sometimes
even in nearly opposite courses. In West Galway
there are various systems of striæ, some going westward, others south-east or north-west, and on one
rock-surface two or three different movements of the
ice are recorded.[1] The newer striæ must have been
engraved by an ice-stream of considerable magnitude,
which probably existed for a long time. If a glacier
is capable of much denudation, all previous rock
markings, especially such small ones as striæ not
the tenth of an inch in depth, should be obliterated
by it. Nevertheless, this is not the case, but in
some localities the oldest striæ are so distinct that
the relative ages of striæ on one rock-surface might
be questionable, were it not that in other places
their true ages were proved. It may be asked, If

---

[1] "General Glaciation of Iar-Connaught and its Neighbourhood,
with a Map." By G. H. Kinahan and M. H. Close. Hodges & Foster,
Dublin.

E

a glacier, after working for years, cannot grind out
scratches not the tenth of an inch in depth, how has
it been able to form the *roches moutonnées* or ice-
dressed hummocks ? These, however, were principally formed when the ice first occupied the country
and the action of the glacier was at its maximum,
carrying away the weathered and softened portions of
the rocks. If we examine a weathered mass of rock,
we find that the atmosphere has affected it more or
less deeply, the lower margin of the weathered portion being a line made up of a series of curves that are
abrupt, if there are sharp inequalities on the surface
of the rock, but are gentle if the surface is even or
nearly so.   If an abrader such as ice were to attack
such surfaces, the weathered portion would be quickly
removed and leave a flowing outline similar to that
of an ice-dressed hummock. A second ice, passing
over such a surface, would have no weathered portions
to remove; consequently it could not abrade the
surface like its predecessor, or obliterate all the
scratches and ruts produced during the passage of
the first sheet of ice.

A glacier when it first occupies a country is charged
with the detritus due to weathering, and if it dies
out previous to all this contained matter being
carried away, the country it occupied will be
covered by hummocks of rocky moraine drift, as
proved by King in his explorations of the dying out

glaciers on the West Pacific slopes.[1]   If, however,
a glacier traverses a country already cleared by ice
action, this ought not to be the case, for (similarly
to marine action acting on glaciated rocks), all the
loosened rocks would have been removed previously,
and the new force would have to act on a hardened,
planed, and more or less smooth surface ; consequently
there would be few or no inequalities for it to act on,
no debris for it to gather up, and few tools with
which it could abrade and triturate the rocks.

In West Galway large tracts of ice-dressed country
were noted, in which there is scarcely any drift and
only a few erratic blocks.   In some places marine
denudation may have removed the drift since the ice
melted away. This may be possible when the country
is below the level of the margin of the " Esker Sea "
(350 feet above the present sea level) ; yet it is not
probable, as the hummocks of rock not only give
evidence of having been planed and grooved, but also
in places still retain their etched and polished surface;
and although the groovings might survive the abrading of sea-tossed blocks, yet the polishing and etching
could scarcely escape.   Furthermore, in other places
the ground is so high that marine action could not
have reached it during the " Esker period ; " and it

---

[1] " On the Discovery of Actual Glaciers on the Mountains of the
Pacific Slope."   By Clarence King, U.S. geologist.   *The American
Journal of Science and Art.*   Third Series.   Vol. i. pp. 157, *et seq.*

is evident that previous to that time, the last occupier of the hills was ice; consequently we are forced to believe that in such elevated situation the glaciers died out, leaving little or no rock detritus after them.

At one time we put much more faith in the abrading powers of ice than we now do, but experience teaches that first impressions are sometimes deceptive, and that often the more a subject is studied the more apparent becomes the necessity of modifying first ideas. Undoubtedly, under favourable circumstances, ice must be capable of executing a considerable amount of work. This is also true of all the other denudants, but none of them can operate with advantage unless it has received aid from its fellow-workers. If ice, as long as it continued moving, had the same denuding powers, we could have no such thing as an empty rock basin, as the ice when it finally melted would be full of blocks and debris, and would more or less fill up the hollow. If, however, ice has less opportunities of denuding the longer it works, we might in time have an ice pretty nearly pure, so that when it finally melted there would only be a mere trace of drift left by it; and this seems to have been the case in many of the rock basins in Iar-Connaught, as the only deposit in them is a peaty one of a quite recent age.

CHAPTER VI.

THE DENUDANTS, OR CARVERS OF THE EARTH'S
SURFACE—METEORIC ABRASION.

In the last two chapters it has been shown that neither the sea nor ice have much power as denudants unless aided by meteoric abrasion. Meteoric abrasion seems to be the most universal performer in the great work of denudation. Besides, it is not merely a surface-worker, as its influence can be found at a considerable depth below the surface. But extreme subaerialists would make their favourite agencies do an enormous quantity of work; and to prove these theories they base their calculations on the fallacy, "that all the surface of the ground is being more or less denuded." This, however, is contrary to fact. In tropical regions there must be an enormous amount of denudation, as the hot winds and sun parch up the surface, forming great thicknesses of debris, to be carried away by torrents during the subsequent rains. Still, even in those climes, denudation does not affect the whole area, as miles are covered with jungle and forest, sure protectors of the surface. In alpine and arctic regions the everlasting snow and ice act in a similar manner, and it is questionable if, in the temperate regions, meteoric abrasion is able to gradually denude even half the surface exposed; for in damp districts, undoubtedly, *it would naturally act more as a preserver than a destroyer.* This is well exemplified in many mountainous mining districts; for in such localities fuel is often scarce, and the miners cut, for firing, the peaty surface soil that has clothed the mountain and preserved it from denudation; consequently the rain and other agencies act upon it for a season, cutting the surface into small ridges and hollows. Nevertheless this does not last long, as the dampness of the climate soon clothes it afresh with a peaty garment, and thereby effectually preserves it. In a district occupied by miners, all the different stages can be observed : the places from which turf has not yet been cut, where there has been no denudation; the spots newly bereft of turf, where denudation is at work; and the old turf-bogs where the mountain is again covered with a thin coat of peat.

A group of table-topped hills, capped by peat and bounded by steep escarpments, is a good illustration, showing the contrast between the destroying and preserving powers of meteoric influences. First the

original mass, by denudation of some kind or another,
would seem to have been planed into a flowing,
gradually undulating table land, with gently sloping
sides.   Subsequently, by one or other of the kinds of
denudation, or, what is more probable, by two or more
combined, valleys and ravines were cut traversing
the table land in various directions, and forming the
individual hills.   When meteoric abrasion began to
work by itself, or a short time afterwards, peat must
have begun to grow on the hill-tops, which effectually
prevented them from being denuded; therefore those
powers could only act in the valleys and ravines, on
the sides of which they formed steep slopes, crowned
by perpendicular escarpments usually a few feet
higher than the thickness of the envelope of peat.
The deposit under the peat is generally the remains
of the original surface drift of the ancient undulating
table-land, and as the covering of peat now protects
its surface, meteoric abrasion works in sideways,
undermining the peat, which consequently topples
over in large pieces when its weight opens the joints.
As the waste thus proceeds both vertically and hori-
zontally, the area of the flat summits must decrease,
although most slowly.   Favourable circumstances
will much accelerate the horizontal denudation,
such as in some districts continuous and prevalent
winds from certain points of the compass; these
rapidly undermine the peat; consequently on such

sides of the hills the slopes are much more gradual than on those sheltered from the wind.

In cultivated districts, where the valleys are inhabited and tilled, the protected area of the whole surface must gradually diminish; for, as the valleys widen, the quantity of tillage is increased. But in wild districts, where nature still reigns, what is taken from the protected summit has an equivalent returned in the valleys; for when the latter have become wide enough, peat forms there also, which, once it has established itself, rapidly grows not only on the flat, but also creeps up the slopes; so that eventually there is only a small fringe round the edge of the table-topped summit which can be acted on by the denuding forces.

In moist climates, not only on mountains, but also on low lands (if uncultivated), unless drained naturally or artificially, peat will grow and defy denudation. Even in cultivated tracts, when the land is under grass and not in tillage, denudation can have very little effect. This is well illustrated when a section is opened down a hill-slope that has been for a long time grass land, more especially if it merges at the bottom into a boggy flat. In this instance it will be found that there is scarcely any difference between the thickness of the surface soil at the summit and base of the hill; while the peat from the flat has been gradually encroaching on the

grass land and creeping up the hill. On the other
hand, if a section is cut down a hill similarly circumstanced, except that it has been tilled for a considerable time, the results will be different, and the
denudation quite apparent; for the surface soil will
be much reduced in thickness on the slope and
summit, it being carried down the hill, and even
found overlapping and lying on the peat in the flat
below. Such a section will also give a record of the
number of times and periods the hill has been in
tillage and under grass; for the peat and surface soil
at the base of the slope will be found intercalated
according to the number and length of the times the
hill was exposed to the denuding influences. It may
be said, that although meteoric abrasion does not
denude the surface of the grass land, yet the water
percolating through the soil increases its depth by
gradually rotting the subsoil. This also seems to
be disproved, as it is not of uncommon occurrence,
when draining grass land formerly tilled, to find
lying on the surface of the subsoil the old sods.
This could not be the case if the atmospheric influences
had the power of increasing underneath the surface
soil of grass land. Different authorities have shown
that in some localities "stones grow on the surface;"
that is, the surface of some grass land gradually
becomes covered with stones. This, however, is only
the exception to the general rule, and in moist climates

can never occur, except in places that are naturally
subjected to extreme drainage; such as chalk hills, a
country over cavernous limestone, a porous sandstone,
or the like. Calcareous rocks are, in a great measure,
chemically formed, and are more susceptible than
other rocks; consequently they more readily suc-
cumb to meteoric abrasion, the carbonate of lime
being carried off in solution, while only the insoluble
parts, such as flint, chert, and clayey matter remain.
Besides, during warm weather such lands will be
scorched and burned by the sun, thus forming dust
to be carried away by wind or rain.

Engineers are well aware of the protecting power
of grass, and they grow it on the slopes of their
railway cuttings and embankments, by which means
all denudation is stopped in two or three years;
after which the surface is added to by the growth
and decay of vegetation, and the work of earthworms
and other soil-producing animals.

Meteoric abrasion can perform a vast amount of
work in tropical climates, as pointed out in South
America by Agassiz, and in Abyssinia by Blandford
in his sketch of its geology. It should, however, be
remembered that meteoric abrasion is most active
in such climates, and that these, in by-gone ages,
could only have existed on limited portions of the
surface of the earth: yet, if we look back into the
records of the past, we find among the Laurentian,

the Huronian or Cambrian, the Cambro-Silurian, the Silurian, the Carboniferous, and many of the newer formations, that conglomerates and such accumulations are not uncommon; while in places they are so massive and so frequent as to be characteristic of geological groups.   Meteoric abrasion could not have formed such rocks ; glacial action may have had something to do with them ; but prior to their being deposited as we now find them, the fragments and blocks of which they are formed must have been tossed and rolled about by the sea or some other large accumulation of water. At the present day, in places on the coast of our own little islands, we can study deposits accumulating, that are identical in character with such rocks. In the tropics the principal rocks that are being formed by meteoric abrasion, are made up of more or less fine materials, disintegrated by extreme heat, and subsequently carried away by wind, or rain and rivers, to be deposited in lakes or seas.

If we look into the present conditions of Ireland it is apparent that the growth of soil, except on the coast-lines, must equal, if it does not exceed, the denudation by meteoric denudation.   On the upper portions of all but a few of the limestone hills, an envelope of peat is growing which is yearly increasing, while two-thirds of the lowlands are in a similar condition. Of the rest of the country a half at least is under grass-land or wood; consequently very little denudation can

take place, except the small erosion that is effected
on the banks of streams, rivers, and lakes.   Furthermore, most of the detritus carried down from the
uplands lodges on the lowlands, so that only a small
portion of the particles denuded on lands bordering
the coast-line, can be carried out and deposited in the
sea.   It may be also pointed out that the present
conditions of this island are not more favourable
than those last preceding; as it is known, from
geological and historical records, that in ages past
the country was a vast forest.

The principal work of meteoric abrasion is to
disintegrate rock surfaces, thus preparing matter to
be carried away by wind, rain, rivers, ice, and the sea.
Sometimes, however, it quarries out masses, as
alternations of heat and cold will open cracks and
joints in rocks.   These may become filled with wedges
of water or ice, and such wedges will shift and displace masses of matter ; but, strictly speaking, such a
denudation is not solely due to meteoric abrasion, as
without the cracks resulting from the contraction of
the rocks the denudant could not have quarried the
rocks.   Our former colleague, Prof. T. M. Hughes,
of Cambridge, thus describes the denudation of the
cliffs at Sheppy, Kent, England :—

" In the hot weather the surface of the ground is
cracked from the shrinkage of the clay, and the
cracks are seen along the highest ground as gaping

fissures about six inches across, and many yards in length and depth. When the rain falls, it finds its way into these, and softens the clay at the bottom. If the crack is near the cliff the half detached mass slips down to the shore below." [1] Mr A. Geikie, F.R.S., in his description of the formation of a valley or ravine, has to acknowledge the necessity of the shrinkage fissures ; for while writing of the action of springs, he states that the water percolates through " the joints and fissures of the rocks." Similarly, when describing frost action, the necessary assistance of joints and fissures is mentioned ; and in the description of the denudation of a rock to form a cliff, there is the following :—" There is no long escarpment or cliff of such materials which is not traversed with joints, faults, or other divisional planes, that serve as subterranean channels for the water." [2] So also has our colleague, Mr J. C. Ward, when writing of the scenery of the Lake District, England; as he says of the weathering of the hills over Derwentwater :— " Numerous irregular joints and fissures assist in the work, and hence the craggy hill-sides."

The sun opens cracks and other shrinkage fissures; and without these, as we have pointed out so often, none of the denudants can work; yet all breaks of

---

[1] "Mem. Geol. Survey." The Geology of the London Basin. Part I., p. 387.

[2] "Scenery of Scotland," pp. 35–37.

this class cannot be placed to the sun's credit, as some may be due to heat that has its source beneath the surface of the earth, while many are due to the movements in the earth's crust.

If the sun acts alone or only in conjunction with wind, it is a destroyer; but when it is joined with cold and rain, the effects are modified, as all combined, cause vegetation which eventually clothes the surface of the ground with a protecting envelope. Thus it is evident that the sun ought to accomplish its maximum amount of denudation in tropical regions; yet the amount of work done in the temperate zone must be considerable, if we may judge from what can be observed in Ireland, as the part the sun has taken in forming its present features is conspicuous. All the hills in that island, with very few exceptions, have much longer slopes towards the south than towards the north. These may have gradual slopes southward, while northward the cliffs are more or less precipitous; and with scarcely an exception, all the streams in a mountain group that flow southward have their source and watershed close to the summit of the escarpment that bounds the high lands on the north, so that respectively in all mountain groups the streams that flow southward are much more considerable than the streams that flow northward. This is apparently due to the rays of the sun acting as a denudant on those parts of the land that are most

under its influence, the rock alternately expanding
and contracting, and thereby disintegrating more
rapidly there than in the land that has a northern
aspect, where the changes in temperature are less.
The different degrees of weathering due to aspect are
very apparent during frost; for at mid-day, if a
bank or cliff looks south, and there is a strong sun,
numerous blocks and fragments will be loosened and
drop away, while in a bank or cliff only a few yards
distant, but having a northern aspect, not even a
fragment will be displaced. This may be studied in
artificial cuttings, such as those for roads—the north
bank yearly weathering away, while years must
elapse before the south side loses the form given it
by the workmen. The effects of the sun's rays may
also be traced in a bog which, during dry weather,
cracks with the heat, as in general the principal
cracks are nearly perpendicular to the sun's rays a
little before noon.

The sun's heat may have opposite effects under
different circumstances. Generally, heat will expand
the matter on which it acts. This, however, may be
sometimes not the case with rocks, as the component
strata of the earth's crust contain more or less
moisture, which the sun's heat evaporates, thus
causing the rocks to contract; but at the same time
there would be at the first an expansion due to the
heating of the moisture prior to its evaporation.

These operations, combined with those of cold, can break up the hardest rocks into minute particles.

Cold acts both as a destroyer and a preserver. If matter is dry, it will, as an almost universal rule, contract by cold; if, however, it is wet, the moisture by which it is permeated will freeze and expand. Cracks opened by heat may be filled with rain or moisture, cold will freeze and expand the water, and thus widen the cracks. On the other hand, cold will change rain into snow, the latter will form an envelope on the ground, this while it lasts will prevent meteoric abrasion.

Wind is a denudant of no mean power. It works differently from most of the others, for while they, in general, carry the debris they form from higher to lower levels, wind may carry it anywhere, even on to higher ground. The effects of wind on a perpendicular cliff are unique. Heat, frost, rain, or the sea, acting on a cliff, will, with rare exceptions, form a talus at the base, consisting of the detritus that is constantly falling down. The sea eventually may indeed carry away the debris; but while it is doing so the upper portion of the cliff will probably be weathered backward by meteoric denudation, and it loses the perpendicular character; while heat, frost, and rain, will change an inland cliff into a slope, unless there is a lake or river at the bottom of it. The character of wind denudation, however, is to act evenly on all parts of a cliff, for as fast as the denudation goes on, the debris is blown away often over the top of the cliff, therefore a perpendicular cliff denuded by wind ought generally to retain its vertical character.

In Ireland, wind seems to preserve, in a great measure, the vertical fronts of the huge step-like cliffs or terraces, for which the barony of Burren, Co. Clare, is remarkable. Its effect, however, would be small but for the help received from heat, cold, and rain, they disintegrating the rocks and forming a material easily carried away by wind. Chemical action must also materially assist by dissolving away large portions of the rock (limestone). The rain in this district acts peculiarly, as most, if not all, of the matter it removes is carried off in solution, little or no accumulation of surface soil taking place, while the joints and cracks remain as open fissures. The wind blowing through these fissures tends to carry all the insoluble residue away; but it cannot always accomplish this, as, during the summer, plants grow in many of the cracks, and their foliage prevents the wind acting.

Wind work can be studied in soft sandstone cliffs, such as those of New Red Sandstone age on the coast of Devonshire, where, after a gale, the direction of the wind can be traced on the face of the cliffs, it forming a miniature " crag and tail " (if it may be so

F

called), due to hard particles or pebbles in the rock, that break the force of the blast and preserve small portions behind each—the latter tailing off from the pebbles. Similar wind markings have been observed on the sides of cuttings in peat, sand, and even boulder-clay drift, but in the latter it cannot be as regular or well marked, on account of the varied nature of the materials forming the mass ; but, in most of these cases, the wind action may be assisted by the pelting of rain drops. Even in the older rocks of a much harder nature, conspicuous wind work has been noted, such as in the slates on the west coast of Cork and Kerry, where the cleavage planes are opened, and the smooth surfaces transformed into a system of knife-edges. Here, however, the wind is aided by water and heat, the latter opening the cleavage planes through which the water is driven by the wind. The sea, also, must aid the work, as this peculiar weathering was only observed in rocks within reach of its spray.

Travellers in tropical regions have remarked the weathering out of the joints and cracks in the rocks, by which are carved out fantastic pillar-like masses, or piles like massive cyclopean masonry. This weathering appears nearly solely due to the combined action of heat and wind, aided in some cases by rain ; the first opening out the joints and gradually dis-integrating the adjoining portions of the rocks,

while the wind clears out and carries away the
particles as fast as they are formed; thereby
causing what was once solid cliffs of rock to have a
built appearance. That heat and wind unaided can
thus work is proved by weathered rocks occurring
in rainless regions. In many places in our own
climate, but more especially those exposed to sea or
mountain blasts, work somewhat similar, except that
it is on a much smaller scale, can be studied.
Wind work evidently has a great deal to do with the
formation of those peculiar rock masses that have
been called by such names as "mushroom"- and
"table-rocks;" as it drives the rain under them, also
into the crack and joints, to be there contracted and
expanded by heat and cold, which disintegrates the
moistened portions, while subsequently the wind
again acts and carries away the particles. It may be
asked, Why do not these agents act on the whole
surface of the rocks? So they do, but necessarily the
maximum work must be in the immediate vicinity of
joints and other similar lines, opening them and
weathering away the adjoining parts of the rock.
"Mushroom"-rocks are also formed by wind in conjunction with the waters of a lake, when the joints or
other lines have very little effect, as the rocks must be
worn away at heights corresponding with the ordinary
summer and winter levels of the waters. Chemical
action must here assist considerably, especially if the

rocks are limestone. Similar carvings are said to be done by the sea, but we have never seen any examples of the kind, and it would appear necessary, in order to accomplish it, that the sea should be tideless.

If mushroom-rocks have been formed by the aid of the water of a lake or a sea, they will be found on certain horizons, which is not the case when they are due to meteoric action. On the hills of Devon and Cornwall, England, and the crags of Sligo, Mayo, Galway, and other places in Ireland, they are numerous, and it may be seen that, in general, the side open to the prevailing wind is more rapidly weathered than the others. Wind action may also scoop out a soft bed in a cliff, undermining the upper beds. On the Formnamore mountains, south of the Erriff river valley, Co. Mayo, there are, in places, outliers of carboniferous rocks, and in them are subordinate beds of friable sandstone. The latter, when exposed to the south-west, from whence come the most prevalent winds, are rapidly denuded, the wind driving the rain into them, to disintegrate them, and afterwards blowing them away piecemeal, thus undermining large masses of rock which eventually topple over and form a rocky detritus.

The moist climate of West Galway diminishes, by nearly one half, the amount of work the wind can do in this way; as, during wet seasons, deposits of blow

peat and the like fill all minor cracks in the rocks ;
and if there is a succession of wet years, these
accumulations become clothed with vegetation which
protects the joints from denudation. In ordinary
years, during the hot weather, these deposits become
dust and are gradually blown away; but until this is
done, the wind or other denudants cannot act on the
sides and bottoms of the cracks.

In the same county a succession of wet years will
prevent a vast amount of surface denudation by
forming protecting envelopes of peat, and in some
cases a study of their formation is extremely interest-
ing. Many of the ancient marine (?) cliffs have
below them a shingle talus, composed of blocks
ranging from the size of a man's fist to tons in
weight. These tali are probably due to meteoric action
since the land rose, yet many of them are now covered
by a growth of peat, which in places has reached a
depth of eight feet. Any one who has studied the
growth of peat is well aware that it forms and grows
more readily on a non-porous than on a porous substratum ; large portions of Ireland, as for instance
the mountains of Burren, on account of the porous
nature of the underlying rock, being destitute of bogs.
These shingle beaches on this account ought to be
unable to grow peat ; neither could they do so but
for occasional successions of wet years, during
which the wind and rain carries on to them from

above peaty matter, which remains long enough to be
enveloped with verdure, the growth and decay of
which yearly forms peat, which afterwards creeps
downward, till eventually all the shingle is covered.
That such must be the process is apparent in various
places where, as yet, the work is incomplete.

Rain and *rivers* by themselves seem incapable of
doing much work, but when combined with the other
denudants, and with cracks or fissures, their effects
are considerable. The connection with cracks, joints,
and the like, is almost ignored by extreme subaerialists,
yet they inadvertently admit that there is such.[1] An
illustration of the powers of rain and rivers was remarked near Draperstown, Co. Londonderry. There,
one of the streams, a head water of the Moyola river,
has its bed partly in drift and partly in a soft friable
conglomerate. At first it might be considered that
the upper part, in the drift, was solely due to river
action ; but on examination it will be found that this
is not the case, as there is a break in the drift,
immediately north, and partly along which the ravine
has been excavated. This stream, during flood, carries
down innumerable blocks and pebbles which apparently ought to rapidly wear away the conglomerate ;
yet this is not the case, for during the ages it has been
at work only a few feet in depth have been excavated,
except in those places where the bed of the stream and

[1] See page 77.

a break in the strata coincide.   Many other striking examples could be pointed out, especially among the ice-dressed hills of Galway, Kerry, and Cork, where streams have run over polished, scratched, and etched surfaces of rock for ages without having been able to obliterate the ice-marks.

It has been shown already how rain forms, in the west of Ireland, a protecting envelope of peat which stops denudation; and not only there, but also over a large portion of Ireland, the moisture of the atmosphere has a very similar effect; a considerable portion of the country having a greater or less coating of peat.

Chemical action, although an assistant to the other destroyers, may also be, in an indirect way, a preserver, as it disintegrates rocks and forms soil in which plants grow and eventually form a protective envelope.

We have now given short general descriptions of the work capable of being done by the different denudants, and have pointed out that each, individually, can accomplish very little; but if two or more are combined the work may go on much more rapidly, while all work most efficiently when assisted by joints, cracks, and other shrinkage fissures that facilitate the disintegration and quarrying of rock masses.   Hereafter, when describing the various features of the earth's surface, we will have to refer to these different denudants, and occasionally may have to describe their work more minutely.

CHAPTER VII.

THE RELATIONS BETWEEN FAULTS, OPEN JOINTS, AND THE FORMATION OF VALLEYS.

But for the existence of faults, joints, and other shrinkage fissures, very few, if any, valleys could have acquired their present form. Those that could be formed without their aid are valleys due solely to the strata being bent into synclinal curves. Such valleys may exist, but they are of very small extent and of rare occurrence, on account of the operations which have formed the features of the earth's surface. Prof. Le Conte's description of the causes to which are due the elevation of mountains may be thus epitomised.[1] Off a shore line masses of sediment gradually accumulate, and their weight causes a slow subsidence; but when the accumulation of sediment becomes of sufficient thickness there follows a rise of the geo-isotherms, and an invasion of the sediments by the interior heat of the earth. A temperature of 800° is sufficient to produce, not only metamorphism, but

[1] "Formation of the Features of the Earth's Surface." *The American Journal of Science and Art.* Third Series. Vol. iv., pp. 460, *et seq.*

aqueo-igneous pastiness, or even complete aqueo-igneous fusion.   Finally, this softening determines a line of yielding to horizontal pressure; consequently, as the interior of the earth contracts and the shell is crushed up, the upswelling takes place in these lines. The enormous folding of the strata which must occur in the formation of mountains by lateral thrust would, of necessity, produce fractures at right angles to the direction of the thrust, and the walls of such fractures would often slip under the influence of gravity, or else might be pushed one over the other by the sheer force of the horizontal thrust, thus forming the different kinds of faults; the common kind in which the hanging wall has gone down along the foot wall, and the uncommon kind generally found only in strongly folded strata, in which the hanging wall has been pushed upwards.

Such breaks and fractures as those mentioned by Le Conte must form lines of weakness and of shattered strata ready to be removed by the first denudant that may act upon them; while on account of the sheer force of the horizontal thrust, few or no synclinal curves could come up to the surface unbroken, and therefore unmodified, and form valleys solely due to the trough of a synclinal curve.

It may be said that protrusions of irruptive rocks are capable of forming " intercolline spaces " (Lyell), or valleys between protruded hills.   Such hills also, however, would be indirectly due to cracks, as without breaks, through which the matter could reach the surface, we would have no eruptions.

In some mountain groups, such as that of Dublin and Wicklow, some of the valleys have an appearance that at first might lead an observer to imagine that they had been denuded gradually bit by bit; but among other hills, such as those forming the highlands of Scotland, or the mountainous parts of Galway or Mayo, it is generally apparent that the valleys are connected with breaks in the subjacent rocks.

A few valleys seem to have no connection with lateral breaks; such valleys, however, are only minor features, and scarcely deserve the name when compared with the others.

In most straight, or nearly straight, valleys, or in those formed of a series of straight portions, the connection with the associated breaks is easily traced out; but in many tortuous valleys it is not so apparent; nevertheless, after a little examination, it is seldom that proof cannot be found that all the different lines, no matter how irregular, have connections with breaks, either faults or joints. A ravine which illustrates this fact occurs on the east slopes of Slieve Gallion, Co. Londonderry. This glen is most irregular, and after a casual examination its excavations might be supposed to be due solely to the influence of rain and rivers. This, however, is not the case, as the rocks are traversed by systems of joints ; one system (*a, a,* fig. 22. Pl. IV.), parallel to the strike of the beds, while the other (*b, b*) is slightly oblique to it. Of these systems of joint lines the stream (*c, d*) has taken advantage, at one time excavating along the lines of one system, at another time along the lines of the other ; thus a ravine has been cut apparently most irregularly, but really systematically. In all such valleys or ravines the protruding angles will be more or less modified, or even carried away by meteoric abrasion, as in many of the Devonshire valleys, where the relation between the old joints, openings, and the present valleys, is often very much obliterated and obscured.

Valleys in such accumulations as glacial-drift, on account of its irregular composition, are not as uniform as rock-valleys ; nevertheless, in most tracts of country they are inclined to occur in more or less regular systems, as will be seen by examining the geological maps of the central plain of Ireland, essentially a drift country, where, in the different areas, the long narrow bogs, the stream-courses and minor valley-systems, usually have more or less parallel bearings.

In most countries it would seem as if the main valleys were formed in the rocks prior to the drift period, during which they were wholly or partially filled, while subsequently they have been more or less re-excavated. Dana suggests[1] that large rivers may have flowed during all the glacial period in the valleys under the ice-sheets, which appears highly probable, in which case many of the present valleys may never have been occupied by drift. Or even if such rivers did not previously exist, after the ice began to disappear the valley would be kept clear, as, " sooner or later, the water from the melting ice, descending through the crevasses or other openings, would have made streams in all the valleys, even those now dry." " The melting would have gone forward with increasing haste as ·the thickness of the ice lessened ; and all the streams would thereby have been flooded far beyond modern experience, and consequently the work of transportation and deposition would have been vastly accelerated."

In the country lying about the junctions of the Nore, the Barrow, and the Suir (parts of the Cos. Carlow, Kilkenny, Waterford, and Wexford), there are many deep, narrow, and for the most part, " rock-valleys," they having been evidently formed in the rock of the country, their slopes being in nearly all cases meteoric drift, due to subsequent weathering of their cliffs.[2] Glenmore, the valley by which the

---

[1] *American Journal of Science and Arts.* Third Series. Vol. v., p. 200.

[2] When the base of a cliff has been deserted by the sea, it is gradually modified into a slope. Prior to the year 1846, the north portion of the estuary of Wexford Harbour was bounded on the north by

road from Ross to Waterford runs, evidently consisted at first of various straight lines joining into one another ; the top of the southward escarpment still in a great measure showing their old directions, while the bottom of the slopes, joining into the river alluvium, forms more or less regular curves. At each alteration in the bearing of the valley there is a more or less well-marked lateral valley, showing that the changes in direction are due to transverse breaks, probably lines of fault ; but on account of the depth of the meteoric drift this could not be proved ; but in the valleys of the Barrow and associated rivers, such relations are quite apparent.

Various changes in the level of S.E. Ireland have taken place in very recent times, the last having been a rise of from five to ten feet, by which the land acquired its present elevation.

Before this change the tide in the estuaries of the

drift cliffs, averaging from 20 to 40 feet in height, while in or about that year the sea was cut off from them, when the "North mud-lands" were reclaimed. Now (1873) the cliffs have changed more or less into slopes. In a few places this has been done artificially for agricultural purposes, in sandy portions rabbits have materially assisted in the work ; but in general the process consists of the formation of small landslips, associated with ordinary meteoric waste. In some places where the drift is a sand and associated with springs, small bays are being formed in the line of escarpment, the sand during winter being carried out in large quantities on to the flat. This work has been accomplished in twenty-five years. Rock cliffs, under ordinary circumstances, will also be modified into slopes ; these, however, may take centuries, instead of years, to undergo appreciable change.

Barrow, the Suir, and the Nore, rose more than it does now, and the bases of the marginal cliffs were awash and gradually being denuded. Since the land has risen, flats or strands have, in most places, accumulated along the sides, so that now it is only in a few places the sea can reach the base of the cliffs; therefore they are gradually losing their former perpendicularity. At the same time meteoric abrasion has not acted on them as extensively as on such as those in Glenmore, which have been long exposed to its action; therefore their geology can be studied and understood. From them we learn that the changes in the direction of the river valleys are due to transverse breaks joining into, or crossing, the breaks along which the valleys ran, and that most of these lines of breaks are also faults or displacements, some being much older than the others. It is quite evident that some of the faults are pre-carboniferous and others post-carboniferous; but whether any of the latter are post-glacial has not as yet been proved, principally on account of the general absence of glacial accumulations from the valleys. It is, however, probable that some of them may be of the latter age, as recent displacements have been proved to be not uncommon in the surface accumulations occurring on the country to the eastward.

Valleys often occur along the axis of an anticlinal curve. This at first may appear remarkable, but on consideration it is evidently a natural position, as when the strata were folded up tension would form open breaks thereabouts. The upper rocks in a synclinal trough may be wedged and compressed one against another, but the upper rocks on an anticlinal are liable to be separated one from another by stretching, and broken up by the joint-lines opening, leaving the rock more or less loosened, and ready to be carried away by any denudant, whether it be a river, ice, or the sea. In some cases, however, there is probably less material carried away than might be supposed ; for if we take a given length of horizontal strata and bend it into an anticlinal curve, the uppermost beds, under some circumstances, must either stretch or break ; and if the latter, a V-shaped valley would form (fig. 14, Pl. II.), prior to any denudation having taken place. The horizontal width of the valley at the top, b, being nearly equal to twice the versed sine, to radius ab, of the angle formed between the bed ab and the horizontal line at the bottom of the valley; (fig. 23, Pl. IV. horizontal width $= 2$ $db = 2$ versed sine A, to radius ab.)

The folding of strata is worthy of study. If the strata are thin and each bed is capable of moving every way, over and under the associated beds, such strata may be bent up sharply without fracture, the only effect, apparently, being a little tension above and compression below on each individual bed. Excellent examples of such crumpled-up beds may be studied on the coast of Kerry, in the neighbourhood of Valentia, but especially in the cliffs bounding the gut north-west of the hill called Knocknadober, where some of the folds are right angles, others inverted acute angles. Strata when folded either regularly or irregularly, under horizontal compression, need have no open fissures, if quite free to slide on each other, as the material in excess in the trough of the synclinal curves would be pressed up on to the anticlinals. In nature, however, the compression is not truly horizontal, the thrust being due to the shrinkage of the interior of the earth, which crushes and breaks up the solid crust, causing not only an horizontal thrust, but also a forcing up of matter from below, so that in many anticlinal curves we find the oldest rocks forming the highest ground, and in such cases it appears not improbable that the higher beds have slid down the lower; or rather more correctly, that the lower beds have been slipped up from under the higher. Some seem to deny that such a slipping could take place, or consider that the slips must be so small that they are nearly unobservable, and could not have affected the present features of the earth's surface. They seem, however, to forget that even if the angle of dip was only a few degrees and the slip a few inches in a mile, that in twenty or thirty miles the slip would be considerable. There is also the lateral contraction to be taken into account, as some of the strata are seen to shrink more than others, and consequently to add to the width of the valleys.

In many places there are long, narrow valleys or gorges, that appear to be in a great measure due to the contraction of the rocks along lines of fault or breaks; these often form marked features. In the mountainous portions of Scotland and Ireland some of these lines of breaks and faults appear to be post-glacial, as the rocks forming their sides are not ice-dressed; besides, no part of the ravines formed by them is occupied by moraine or boulder-clay drifts, and if their line crosses an accumulation of drift, there is either a marked hollow or a displacement of the drift, the latter often being higher on one side than the other. In Ireland this is well exemplified in the counties of Tyrone and Londonderry, where in numerous places the original sheet of boulder-clay drift, or Till, as it is called in Ulster, has been broken up by recent breaks, generally faults; the original fissures, however, have been subsequently modified, in some cases by water, probably during the "Esker sea" period, but in others they seem only to have been affected by meteoric action.

In West Galway and Mayo, narrow fissures running along lines of faults and breaks form conspicuous features in places not only among the hills,

G

but also in the low country; a marked example extending from Clifden to Cleggan Bay, this being in places over 100 feet deep, and not more than 30 feet wide.

Deep *maums*, or connecting mountain-gaps or passes, frequently cross the hills; two remarkable ones being those of Salrock and Maumturk, the latter giving its name to the mountain range which it crosses.[1]

In the counties of Kerry and Cork are long, straight, narrow, more or less regular and deep fissures traversing the hills, which give a marked character to the country. Some of these evidently are pre-glacial, as the rocks bounding them are dressed, grooved, polished, and etched; but others, such as those stretching across the hills in the neighbourhood of Kilmakillogue Harbour, appear to be post-glacial, while good examples of some of the ice-dressed fissures occur in the vicinity of Mount Gabriel, Schull, Cork. Very large examples of these fissures occur, in places cutting down to the base of mountain ranges, and often diverting the water from one longitudinal valley into another. These, however, will hereafter be mentioned when speaking of the valleys of S.W. Ireland, and at the present we will only draw atten-

---

[1] In the Salrock Pass there seems to have been a fault long prior to the glacial period; while subsequent to it there appears to have been another movement along the old line of break.

tion to the passes through the ridge south of the valley of the Bearhaven mines, and Ballydonegan Bay, the ridge having an average height of about 750 feet, while the nearly level floors of the passes are about 30 feet above the sea, and a parallel pass, a little farther west, is occupied by the strait that separates Dursey Island from the mainland.

If valleys are not connected with breaks in the underlying rocks, how is it that they occur, in regular systems, over large tracts of country? Examine any, but especially a contoured map of Ireland, and it will be seen that the outlines—river-valleys, lake-basins, and bays—occur in systems, the general bearing of which may be indicated by lines. If such systems are not caused by breaks in the subjacent rocks, they must be due to chance, an alternative that even the most sceptical among the subaerialists could scarcely insist on. From the map of Ireland, it will be learned that the most conspicuous system is an east and west one, that forms nearly parallel features; two lines stretching across the island, respectively from Galway to Dublin Bay, and from Clew to Dundalk Bays; while to the north are other nearly parallel breaks that only extend eastward as far as the nearly north-and-south valley in which Lough Neagh is situated; while to the southward a break runs along the valley of the Lower Shannon, and from that to the nearly north-and-south valley occupied by the Barrow; it can even be traced as far as Arklow, but the eastern extension does not form a conspicuous feature. Further south, from Dingle Bay to Dungarvan, is another valley, that of the Blackwater, along which, as far east as Mallow, a large fault is conspicuous.

Between these systems there are others that run in more or less parallel lines. In some these lines have an easting from north, in others a westing, while between these there may be other systems extending more or less east and west. In a few places, especially if the rocks are ancient or metamorphosed, such as to the south-east (Wicklow and Wexford), to the north-west (Galway and Mayo, also in Donegal), and to the north-east, there are limited tracts having minor systems peculiar to themselves; however, the principal lines in the last named extend into the south-west part of Scotland. The connection between the joints, breaks, valleys, and lake-basins of West Galway will be described hereafter in detail, as also those of S.W. Cork, where, although the rocks are not of very ancient date (Carboniferous), there are special systems of master joints, breaks, and faults.

North of the fault (Blackwater valley fault) that has been proved in the valley that extends from Dingle Bay to Dungarvan Harbour, ten large faults, or systems of faults, have been proved and traced, all of which are connected with more or less marked features in the ground, and have a general parallelism to one another—1st, The Slievenamuck fault, bounding the hills, so named, on the N.N.W.; this crosses the slope of the hill, and in general does not form a very marked feature; 2d, The Knockfearna fault, on the N.N.W. of the range of hills of which this peak is the highest summit; 3d, A fault in the valley of the Shannon, that has been traced from Bunratty to Silvermines, and probably extends to Roscrea; 4th, The Formoyle fault, connected with the valley in the centre of Slieve Bernagh; 5th, The Scarriff valley faults; 6th, The Cloonnagro and Corra valley fault; 7th, The Lough Atorick valley fault; 8th, The Derrybrien valley fault; 9th, The Owenaglanna and Boleyneendorrish valley fault; and 10th, The Dalystown river valley fault.

In the valley between Dingle and Dungarvan, a nearly east-and-west fault has been proved from Dingle Bay to Mallow, and a branch of it probably extends to Dungarvan, but the main fault from Mallow goes nearly E.N.E. past Mitchelstown; this part is nearly parallel to the faults just enumerated. It is a downthrow to the northward, as are also Nos. 1, 2, and 3, while Nos. 4, 5, 6, 7, and 8 are downthrows to the southward, but 9 and 10 have downward throws to the northward. These ten faults are of Post-Carboniferous age, and on account of their parallelism, they, and others that could not be worked out, probably belong to one system, and were formed simultaneously. Nos. 5 to 10 are connected with the group of hills called Slieve Aughta, and the relations between them and the features of the ground have been carefully worked out, as will appear from the following :—

In Slieve Aughta there is only one large north-and-south valley, that of Lough Graney; but the east-and-west valleys are numerous, and lines of fault have been proved to occur in each. The latter valleys can be traced from the limestone country on the west to the limestone country at Lough Derg, while the faults could only be proved to have shifted the beds in parts of them.

The series of faults seen in this district are made much clearer by the presence of different kinds of rocks now in juxtaposition, although they were originally several hundred feet asunder; the dark-blue limestone and the yellow and red Old Red rocks occurring on one side of the fault, while the gray and green Silurian rocks are found on the other.

In places, some of what are here considered faults might possibly only be Silurian cliffs, at the base of which the Old Red sandstone and limestones were deposited, as the rocks strike with the line of fault; this is more especially the case with the *Cloonnagro and Corra fault.* But that they are really lines of dislocation seems probable, as there are so many of them in nearly parallel lines ; also the basal beds of the Old Red on the upthrow and downthrow sides are similar, which would scarcely be the case if they were lines of cliffs.

Bounding the south of Slieve Aughta is the Scarriff valley, which separates it from the mountain group called Slieve Bernagh, which lies to the south. This valley has at least one accompanying fault which bounds it on the north, and seems to run from Feakle Lower, to Mount Shannon. The limestone and Silurians are brought together a little east of a hamlet called Coolcoosaun. A mile and a half S.E. of Feakle the most westerly traces of this fault were remarked. From this it can be traced to Coolcoosaun. On the east it seems to extend to Lough Derg, but no positive proofs were seen farther than the Bow River.

The *second* valley can be traced from Lough Blarnagh to Lough Ea, where it crosses into the Lough Graney basin, and down the Lough Ea or Cahir River to Lough Graney, from whence it runs along the Corra and Derrygoolin Rivers, coming out of the Lough Graney basin on the north of Ardeven. Its fault has been traced from near Maghera Lough to Ardeven—*The Cloonnagro and Corra fault* above mentioned. From the Maghera River towards the

104 THE RELATIONS BETWEEN FAULTS, OPEN JOINTS,

east this fault can be seen in various places, especially at the village of Cloonnagro. It could not be traced across the Lough Graney valley, as the break of this valley probably shifts it, but it occurs again in the Corra River valley.

A *third* valley runs from the road from Gort to Tulla, along the Hollymount River to Loughnagilkagh, on the east of which it crosses the watershed into the catchment basin of Lough Graney, and proceeds down the Drumandoora River to the Lough Graney valley, and from thence along the Bleach and Woodford Rivers to Woodford, crossing out of the Lough Graney basin a mile and a half N.E. of Lough Atorick. Its accompanying fault lies a little to the north, and has been proved from near Carheeny Lough, by Lannaght, to Corlea Bridge, and from where the valley ends at Woodford to Lough Derg; but between Corlea and Woodford it was not remarked, as the country is covered with either bog or drift. At Corlea the Carboniferous limestone is brought down against the Silurians. Near Woodford the limestone and sandstone are thrown against one another. On the shores of Lough Derg, near the house called *The Lodge*, four miles east of Woodford, this fault brings down the unstratified against the stratified portion of the Lower Limestone. From Corlea to Carheeny the fault is well marked, dividing into two branches on the west of the village of Lannaght.

From this valley, near Lough Atorick, there seems to be a branch fault running toward the east, north of the hill called Tullymore, past Drummin and Boleymore, into the bog bounding Lough Derg at Spa Island. This fault could not positively be proved, but it seems to shift the rock N.W. of Oghilly House considerably towards the east, while the wells near the shore of Lough Derg are probably due to it.

The *fourth* of these valleys extends from Lough Cooter, along the Owendalulleegh River, to Marble Hill. In this a fault has been traced from Lough Cooter to near Marble Hill. We have called this *The Derrybrien fault*, as it seems to have its greatest throw in Derrybrien valley, immediately south of the hamlet called Derrylaur, eight miles east of Gort, where it brings down the limestone against the Silurians. In the same valley, at Chevy Chase, five miles S.E. of Gort, its throw is also considerable, as there the limestone is now in juxtaposition either with the Silurian or the basal beds of the Old Red sandstone.

A *fifth* valley extends across the mountain group north of the summit called Cashlaundrumlahan, with which a fault is connected. We find in the Owenaglanna River valley Lower Limestone shale, whose south outcrop could not anywhere be found; these rocks probably are bounded on the south by the S.W.

extension of this fault, as farther west, in the Boleyneendorrish River valley we find an outlier of Lower Limestone shale, which has been proved to be bounded on the south by a fault; but the country on the south of the Owenaglanna River is covered with a deep drift or bog, and no positive proof of the westerly extension of the fault was obtained.

The *sixth* fault in connection with Slieve Aughta was proved across the low country to the N.E. of Dalystown, and at the eastern end of the Dalystown River valley. This, a little on the north of the village of Tynagh, brings up a tract of Old Red sandstone, which forms low hills that are well marked, looking at them from the south. This fault has been proved as far as Hearnsbrook on the N.E., and Dalystown on the S.W., but it probably extends much further in the latter direction.[1]

A remarkable north-and-south feature, that is evidently connected with a fault, is the valley of the Barrow, in the S.E. of Ireland; another, having a similar direction, is the valley from Newry to Antrim, in the N.E. of Ireland; while there appears to be also another, although not so well marked, parallel thereto, extending from Youghal to Thurles, in the south of the island. These three lines of break seem to be of comparatively recent age, as they cut through

---

[1] Memoir and Maps, Geol. Survey of Ireland, sheets 115, 116, 124, 125, and 134.

the systems of breaks with which they come in contact, and shift them more or less.

It may also be pointed out that the general bearings of the lines of coast are all more or less connected with the systems of lines in the adjoining portions of the island, as if the lines that induced the valleys also had an influence on the coast features.

CHAPTER VIII

In different mountainous tracts, the relations between faults, breaks, and lake-basins could be pointed out; but at present we will confine ourselves to West Galway, or Iarconnaught, on account of the opportunities of observation there afforded, large tracts of the country being nearly bare rock, or covered with a thin coat of peat.

In this area lakes are excessively numerous, especially in the low country between Clifden and Roundstone; they are also variable in character, yet a general classification of them may be made under three heads, namely—1st, Bog-basins; 2d, Drift-basins; 3d, Rock-basins.

Bog-basins are surrounded by peat, and are locally called Loughauns or Ponds, and in general have no surface outlet from them. They are connected with breaks in the underlying strata, as all are supplied by springs, to which they are due. The following must have been the mode of growth:—As the peat was gradually developed in the country, it could not grow over these springs, but formed a wall round

them, pounding up the water, and thereby giving rise
to a small lake.   When a loughaun is situated on
low ground, there usually is a swamp or morass connected with it; but if on high ground, or on a watershed, as is not uncommon, its margin is well defined,
and generally higher than the surrounding bog.   This
high, prominent margin seems due to the spring-
water equalising the temperature during the different
seasons of the year; therefore, in the frosts of winter
and the extreme heat of summer, when vegetation is
stopped on the rest of the bog, here it is in full
vigour.   Usually the springs are perennial, and the
sides of the ponds are perpendicular, or may even
cove in; some, however, are only full in winter or
wet weather, consequently in summer the bog at the
sides weathers and falls in, making the loughaun
more of the nature of a boggy pool.   The drainage
from a loughaun is nearly always by soakage through
the surrounding peat; sometimes, however, it must
be subterranean, probably between the peat and the
underlying strata, while in a few cases a crack has
opened during a dry summer in the surrounding bog,
and been kept open by the stream from the loughaun
flowing through it.   From the above description, it
is evident that bog-basins are not due to denudation,
but to a growth of vegetable matter, while indirectly
they are caused by a break or fault in the subjacent
rocks from which the spring issues.

Drift-basins may be situated in hollows in drift accumulations, but most of those remarked in Iarconnaught are partly formed of rock and partly of drift; the drift occurring as a mass or bar across a hollow in the rocks, and thereby pounding-in the water. Some drift-basins situated in low valleys, such as the basin of Kylemore Lake, appear due to bars and mounds of sand and gravel that extend across, or accumulate in, a valley; the ridge and mounds apparently being ancient sea-work. The formation of such gravel accumulation will hereafter be more minutely described. Another kind of true drift-basin is instanced by that of Bealawaum Lake in the corry north-east of the summit of Mweelrea; the hollow apparently having been formed, in a great measure, at the same time as the drift, as it is probable that such irregular accumulations of drift are the debris contained in the ice when it finally disappeared.

For the formation of basins partly rock and partly drift three agents may be suggested, viz., marine, glacial, and meteoric action.

Marine action may build up a bar across the end of a narrow bay or estuary, into which a stream flows; and such a mound now exists at the east end of Cleggan Bay (Map, fig. 16, Pl. III.) This bar is evidently due to the meeting of the fresh and salt water; having begun as a shoal that gradually grew higher and higher by the addition of debris

PLATE III.

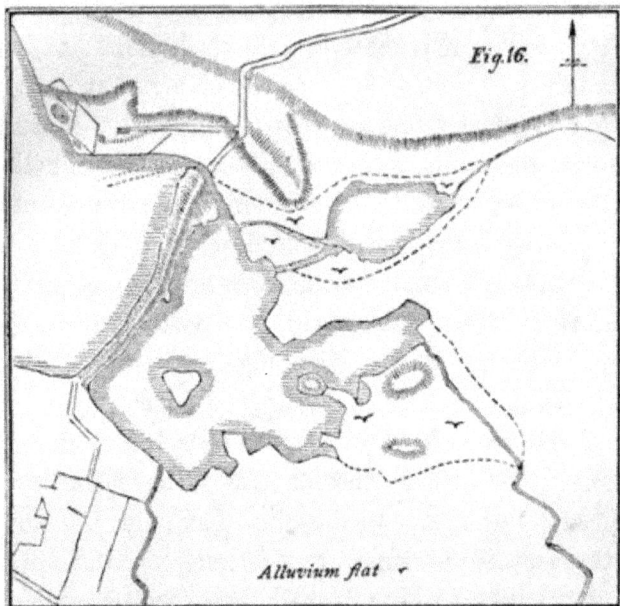

Fig.16.

Alluvium flat

MAP OF BAR AND LAGOON AT EAST END OF CLEGGAN BAY.

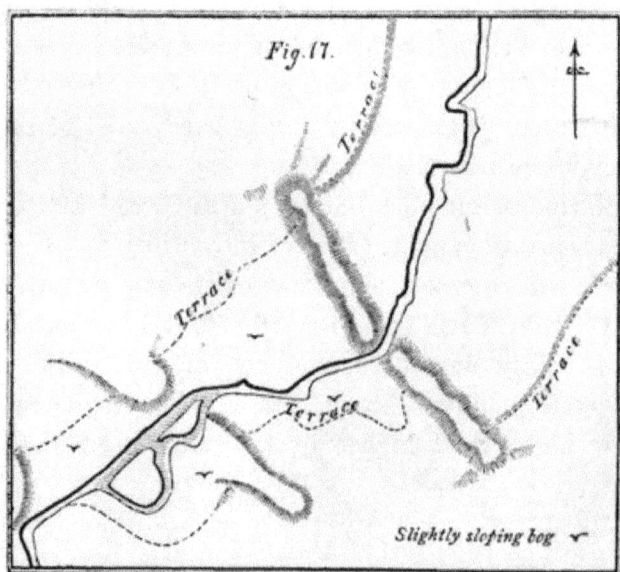

Fig.17.

Terrace

Slightly sloping bog

MAP OF BAR AND FLAT, GLEN CORBET.

driven up on to it by the sea, till eventually it became a mound across the end of the bay, forming
behind it a fresh-water lagoon. The remains and
sites of similar bars and lagoons have been observed
in different places in the neighbouring glens, principally at about the 350 feet contour-line which is
supposed to mark the margin of the sea during the
"Esker sea" period. The relations between the bar
and lagoon at Cleggan, and the remains in the glens,
will appear from their descriptions.

On the north end of the Cleggan bar the sea acts
with the least force, consequently the embouchure of
the lagoon is there situated; naturally the drainage ought to be at or near the centre, where the bay
is deepest, but in that place there is only an under-
drainage when the tide is out. If, however, the land
rose, this drainage, which at present is a soakage
through the sand and gravel, would be continuous, and
gradually increase, till eventually it sapped the bank
and formed an open through it, which the stream
would deepen more and more, till eventually there
was a cut down to the bottom of the gravel mound.
If the gravel had accumulated in a simple bar,
it is probable all the water would be drained out of
the lagoon, and form a more or less level plain; but
if there had been gravel mounds formed in the ground
behind the bar, the bottom of the lagoon would be
uneven; and probably, after the water had cut a new

course, there would be one or more small shallow lakes on the site of the lagoon. That many of the bars or ridges of gravel found in the valleys of Iarconnaught were once similar to the bar at Cleggan is evident, as in addition to the bar, the old stream-course or embouchure is still visible, also the terraces formed at the margin of the water in the lagoon. The accompanying map (Pl. III., fig. 17) represents the site of one of these ancient lagoons, the gravel banks being the remains of the ancient bar; there are also terraces that not only show the limits of the lagoon, but also the sea margin on both sides of the valley outside the bar.

In some bays or estuaries similar bars to that at Cleggan cannot form for various reasons. First, the stream may have greater power than the tide; second, the river may be subject to freshets, and during floods all the sand that had accumulated will be swept away, and spread more or less evenly on the bottom of the bay; and third, the stream may be so inconsiderable, that it cannot form a back-water, between which and the influx of the tide the bar would form. In the latter case a semi-bar sometimes forms, the conditions being unfavourable for its extension across the mouth of the stream, and such a bar occurs as a gravel bank at the village of Leenaun, Killary Harbour.

It may be said that the streams at present flowing into the lagoon at Cleggan are too small to have formed the bar and lagoon. This is probably correct, as it must be borne in mind that at no very remote period all the streams and rivers in Iarconnaught were larger than they are now, as two or three hundred years ago most of this area was a forest. Boussingault, who wrote on this subject, gives facts showing the effects caused by the cutting down of the forests on the streams and springs in South America; furthermore, this fact is so well known on that continent, that some of the South American governments have passed laws regulating the felling of timber.[1]

Other evidence in favour of the Iarconnaught streams having been larger than they now are, is that in different places peat is gradually growing over and filling up old stream-courses.

If a stream debouches on an open seaboard, a bar and lagoon may form if the ground is low and the volume of water discharged is insufficient to sweep away the sand driven up by the wind and sea. Such lagoons were remarked on the coast of S.W. Mayo, but more extensive ones are found in South-east Ireland. These, such as that which existed at Kilmore, County Wexford, are separated from the sea by irregular eskers (*anglice*, ridges) of more or less fine, often stratified, sand. Such eskers are evidently, in the first place, bars formed between the open sea and the shoal water, which afterwards are augmented by sand-

[1] *Edinburgh Phil. Jour.* xxiv. (1836), pp. 89-91, and 102, 103.

H

and-shell debris blown up into them by the wind,
each heavy gale forming a new layer, which the
vegetation binds more or less together, and preserves
from removal by subsequent winds. In some places
the lagoon has become filled with a peaty soil, and
as the south-east of Ireland seems to be gradually
sinking, the sea is nearly imperceptibly moving back
the sand-ridge, so that in places the sand-ridges are
found to have a peaty foundation under them.[1]

Bars and ridges of gravel or sand like those we
have described, can only form near the end of a bay,
or at the margin of the sea. There are, however, other
accumulations, such as the already-mentioned banks
and mounds of gravel and sand in the valley of Kylemore. These would appear to be comparable to the bars
at present forming in straits, or any other place where
two tides or currents meet ; and if the land hereabouts
was lowered 350 feet, the tides would flow and ebb
through the valley of Kylemore, and form similar bars
and mounds to those that at present exist.

That such conditions obtained during the " Esker
sea " period, appear proved by our finding in the
different valleys of Iarconnaught, and also in Clare,

---

[1] Although this coast may be now sinking, it appears not improbable
that a short time since it may have been rising, as part of the ridge
that now margins the mudlands of Wexford Harbour is said to have
been formerly off the mouth of that haven, and was marked on the
charts as dangerous shoals. The estuaries on this coast deserve a much
more careful examination than has yet been given to them.

Limerick, Kerry, Cork, &c., places miles apart,
terraces that mark the margin of this ancient sea.
These gravelly accumulations are generally associated
with rocks to form the lake-basins; and in many
valleys after their elevation, rain and rivulets have
cut channels through the bars and banks, thereby
eventually draining the hollows in which lakes at
the first existed; so that now only in a few, where the
cut was not deep enough to drain all the hollows,
do lakes exist. If we examine the Kylemore valley,
we find that its lake once extended considerably further up and down the valley, and that the small
lakes there at present were only deeps in the ancient
lake, which the present outfall cannot drain.

The third class of bars are those due to glacial
action. Hooker, Forbes, and other Alpine or Arctic
explorers, have described the moraines which form
across valleys and hollows, also the avalanches and
other debacles of rocks, stones, clay, and mud which
slide down into valleys during thaws, damming up
the drainage, and stopping the original water-courses,
often forming lakes of greater or less dimensions.
Snow drifts may act similarly, as the up-side may
become a sheet of ice, and prior to its melting away
a river or stream may cut a new channel, and desert
its old one. In the mountainous parts of Great
Britain and Ireland, bars of drift, apparently of
moraine origin, or accumulations from successive

avalanches, often extend into valleys. Some of these must at one time have stretched across the valleys, ponding up masses of water, and thereby forming lakes. Now, however, in most cases, the stream-courses having gradually become deeper and deeper, the bars have been cut across, and the original drainage of the valley is restored; but not in all cases, as, for instance, Lough Inagh, Connemara, which is separated from Derryclare Lake by masses of moraine drift, to the bottom of which the connecting river has not as yet excavated its channel.

The fourth class of bars are meteoric. Such bars most writers appear to ignore, yet they often form marked features, and their formation can in many places be studied. A bar partly meteoric and partly lacustrine may be formed when a torrent flows into a lake and makes a delta, called in Ireland a *srah*— *i.e.*, the level land at the *inver* or mouth of a river or stream, usually covered during freshets. A srah forms in a wide lake by the waters during floods levelling the surface of the detritus as it is brought down; but in a narrow lake a bar may form across it, and if the original *embouchure* of the lake is after- wards lowered, two sheets of water form, one with a higher level than the other, and in Ireland such a lake is generally called Lough-a-voul or Lough-aw- woul (*anglice*, the lake of the two spots).

Other bars may be due to detritus carried down by

a torrent into a flattish part of the valley. As we have
not seen the work of one of these mountain torrents
described, we may attempt it.   Mountain torrents
are very erratic in their movements : for years they
may flow in one course, but at any time the old
course may be deserted and a new one excavated.   A
newly-excavated torrent-course down a hill-side, has all
the appearance of a more or less deep, artificial trench,
from which the materials excavated had been thrown
in irregular banks or heaps on both sides.   At first
the stream takes possession of some sort of break
in the surface soil, whether a crack due to contraction, a path worn by sheep or cattle, a rut excavated by lightning [1] or the falling of a waterspout,
or a trench cut by man.   This it deepens and widens
at the bottom, till eventually, during sudden floods,
the stream forces up large masses of the surface, and
scatters the debris along the banks of the channel;
or the water may carry some of the lumps along
with it, till caught in one of the narrows of the excavation, when they form a dam.   While the dam exists,
the rush of water from above will force all transported
materials, such as stones, turf, and the like, up on
to the banks on both sides.   Such dams rarely last
long ; but as they are continually forming and burst-

---

[1] The ruts due to the lightning observed on the hills in Iarconnaught had an appearance very like a regular furrow opened by a
plough.

ing in different places along the excavation, the detritus carried down is distributed more or less evenly on each bank, forming piled-up, artificial-looking heaps. Each stoppage also adds to the force of the torrent, as the pent-up water, when a dam bursts, rushes down with increased velocity.

The work done by the same or a similarly circumstanced torrent when it reaches a flat would be different, as there, instead of excavating a channel, it would build up a bar extending out from the hill-side, and in some cases even reaching across a valley, thereby damming up the drainage, and forming a lake. In many cases the streams, for reasons somewhat analogous to those just mentioned when describing the descent of a torrent down a hill-side, will keep on the top of its bar, and thereby increase its length ; but not always, as sometimes it will break over to one side or another; after which, if it breaks over on the upper side, it will begin to denude away the bar.   Remarkable examples of bars of this class occur at the Lake of Geneva, especially on the flanks of the great snow-hills between Clarens and Villeneuve, each bar or mound having the torrent that formed it still flowing along its summit.   Lakes due to the formation of a meteoric bar across a valley are usually shallow, both on account of the form of the ground, and also because the bars are more or less porous ; the sheet of water, therefore, is seldom considerable, except during heavy

rains, when the supply exceeds the drainage by soakage. There is, however, an exception, when one of
these bars forms across the mouth of a corry, a place
in which they are not uncommon, for reasons mentioned
hereafter, while describing the formation of corrys
and their lake-basins.

Other meteoric bars may be due to landslips, as
during dry weather cracks will form in mountain
slopes, whether the drift be meteoric or glacial. If
such cracks run up and down the incline, or perpendicular to the trend of the valley, they only form
channels afterwards to be occupied by streams; but
if they are transverse—that is, rudely parallel to the
lower part of the valley—water, after rain, will collect
in them, and finding no egress, will eventually force
out large masses, that slide down to the bottom of the
valley and dam up the drainage. Hooker has described such debacles in the Himalayas which are of
considerable magnitude; but none of those remarked
in West Galway are very large.

Other bars that have been noted may possibly be
due to avalanches during the time that snow and ice
existed in the hills, as we know that at the present
day, in Alpine regions, avalanches occur in special
places, and by their frequent sliding, accumulate bars
of drift. In other places, lakes may possibly be
formed by beavers' dams. This suggestion, however,
is merely conjectural, and prompted by the form of

some of these lake-basins. Against it is the fact, that although the beaver is known to have been an inhabitant of Wales during the historic period, yet we have been unable to find in the Irish annals any record of its existence.[1]

The great characteristic of the Iarconnaught lakes are their "rock-basins." Such basins Professor Ramsay seems to believe have been formed by ice action, but of this theory Lyell writes, "It appears . . . that the abrading action of ice has formed some mountain tarns and many moraine lakes, but when it is a question of the origin of larger and deeper lakes, . . . it will probably be found that it has played a subordinate part in comparison with those movements by which the changes of level in the earth's crust are gradually brought about."[2]

We, however, would go even further than this eminent geologist, and suggest that, unaided by the other great denudants, and also by cracks, fissures, and faults formed during the movements in the earth's crust, ice is incapable of eroding out rock-basins.

In Iarconnaught, not only is meteoric abrasion

---

[1] There does not appear to be any real native Irish word for beaver. In O'Reilly's Dictionary we find "*Dobhran-leasleathan,* a beaver," but he seems to have taken this from the Scotch Gaelic, as no mention of beaver is known in any Irish MS. *Dobhran* means water-animal, and *leasleathan* broad-tailed. The Welsh for beaver is similar, namely, "*llostlydan,*" broad tail.

[2] "Student's Elements of Geology," p. 164.

going on at the present day, but also, what appears
to be its work in ages long past can be recognised.
Marine denudation has left its records in the raised
gravel-bars and beaches, in the terraces on the hill-
sides, and in the cooms and corrys at similar elevations, not only in the hills of Galway, but also in the
hills of Clare, Limerick, Kerry, Cork, and other mountainous places in Ireland, while at the present day
similar work is being accomplished on the coast-lines.
Glaciers are now doing no work; but that they once
existed is proved by the dressed hummocks and the
planed, grooved, etched, and polished rock surfaces.
It is quite palpable that the rocks of Iarconnaught
have suffered at successive periods from the disturbances due to the movement in the earth's crust, as
they are not only folded and contorted, but also faulted
and displaced to a very remarkable extent. Some
of the faults must be very ancient, as the "fault-
rock" in them was metamorphosed along with the
associated rocks; while others are post-glacial, as was
pointed out by our colleague Mr R. G. Symes, F.G.S.,
as they displace the ice-formed drift. These last-
mentioned breaks are very instructive; for although
they form greater or less hollows, some over 100 feet
deep, and not more than from 10 to 30 feet wide,
yet in general, the horizontal displacement is trifling,
while in some it may be *nil*, as the same strata at
opposite sides of a ravine may face one another, while

the cliffs on both sides may be on a similar or nearly similar level, as if the break had opened or gaped while no vertical or horizontal movement had taken place.

The rock lake-basins of Iarconnaught may be in corrys or maums,[1] or on hill-tops, while others are in valleys, or in comparatively level plains; always a connection between them and one or two lines of master-joints or faults is apparent, which would seem to suggest that these lines of break, whether faults or only shrinkage fissures, must have materially assisted in the formation of rock-basins.

During, and after, the formation of a fissure at the surface of the earth, unprotected either by water, ice, or the like, meteoric abrasion would modify it, forming debris that in part might be carried away by rain, rivers, or wind. Marine action would act nearly similarly, if the fissure was lowered and came under the influence of the sea; so would also ice, if that agent was subsequently formed, or if it existed prior to the development of the fissure;—the traces of the work of these different denudants are preserved in the rock-basins, fissures, gorges, and ravines of Iarconnaught. Another feature of these rock-basins is, that all the long stretches coincide with lines of

[1] *Corry* (coire, *Celtic*, pot or cauldron), a valley having a bowl shape; *maum* (madhm, *Celtic*, the inside part or hollow of the hand), a connecting mountain gap or pass, the equivalent of *col*.

breaks, faults, or other shrinkage fissures, as well as all the transverse guts and bays. This is exemplified in the accompanying map of Lough Conga (Map, fig. 15, Pl. II.), the lake being specially selected on account of its outline being so irregular. All the lake-basins in this country widen or contract in accordance with the number of faults or joint-lines which meet or cross in the area occupied by them. If such lines were numerous, close together, and crossing one another, a considerable portion of the rocks would be broken up, and more or less easily removed, either by ice or sea action; while if only one line or two parallel lines existed along which the force could act, the hollow excavated would be long and narrow. All these cases are exemplified by the Iarconnaught rock lake-basins, some of which are wide, others long and narrow, many most irregular in outline, while all coincide with the associated features of the country due to the breaks and faults in the rocks.

After rocks have been broken up by faults and the like, marine action in a shallow basin could sweep or suck out the broken materials; but in places of considerable depth it appears incapable of executing the necessary work. Meteoric abrasion would take an incredible length of time to do it, as it would first have to disintegrate the blocks and fragments, while the detritus would have to be removed by wind, or through a subterranean passage by water. An ice-sheet, however, appears capable of doing the work more quickly, as it would lift up the blocks and other detritus and carry them off, whether the hollow be situated on a hill-top, in a valley, in a maum, in a corry, or on a flat plain. From this it will appear, that although it is probable both marine action and meteoric abrasion may have accomplished a little work in removing the debris produced by the cracking and faulting of the rocks, yet the chief worker would seem to be ice, which would account for rock-basins being typical of glaciated regions.

After the ice had disappeared from the country such hollows would become filled with water, and form lakes. This, however, would require, in all basins above the sea-level, that the joints and "fault-rock" should be water-tight, or the subterranean vent incapable of carrying off the surface supply, or that drift or other matter should be deposited and staunch the water-basins. It is quite apparent that some of the rock-basins in Iarconnaught have subterranean passages connected with them, while others have had them; but in the latter cases they have since been closed by a peaty deposit carried into the lakes by wind, rain, and streams. There are, however, other rock-basins, that never were, and probably never will be lakes, the fault-rock being so loose and incoherent that no

water can remain in them.   Examples of these
waterless rock-basins occur on the hills of Errisbeg,
near Roundstone and Lettermore, north of Ballynakill Harbour, which respectively are composed of
hornblende-rock and quartz-schist, rocks which, on
account of their hard nature, are not easily broken
small, and consequently are likely to form loose,
open " fault-rock."   In places on the low ground
similar rocks also occur.   Here, however, rock-basins
are liable to be invaded by peaty matter carried
by wind and streams; consequently, in many cases,
basins that doubtless once leaked are now water-
tight.   Nevertheless, in places there are large tracts
solely drained through subterranean passages.

In these rock-basins, especially those on Errisbeg,
the connection between them and breaks or faults
is very apparent.   Some of the latter are evidently
post-glacial, and point out the large vacancies which
may be formed solely by the contraction of the rocks,
and the small quantity of debris that a subsequent
denudant, in such cases, would have to clear out.

CHAPTER IX.

CORRYS WITH OR WITHOUT LAKE-BASINS.

EMBOSOMED in a corry or coom, a small lake is often found. Sometimes in a true rock-basin, but more often in a basin partly rock and partly drift; or even, as previously mentioned (as in the case of the basin of Lough Bellawaum), it may be solely drift. To this subject special attention has been drawn by the Rev. M. H. Close, M.R.I.A., &c., in a paper, " On some Corrys and their Rock-basins in Kerry."[1]

From this memoir it would appear that the author considers the formation of corrys to be largely due to ice action. The following is an epitome of his reasons for coming to this conclusion :—

a. Corrys cannot have been formed by the atmosphere, for this agent is now engaged in trying to soften them away; nor by the sea, for the following reasons : —The sea does not seem to be now producing such features on the S.W. coast of Ireland. The floors of contiguous corrys are often of very different elevations; while there are no apparent terraces on the open mountain-sides corresponding with the corry

[1] *Jour. Roy. Geol. Soc. Ire.* New Series. Vol. ii. p. 236.

floors.   Corrys have no perceptible preference for places which were most exposed to the action of the sea.

b. These hollows are of a different order from the other shapes of the mountains on which they occur, which argues a special mode of formation.

c. The rock-basin, when there is one, is an integral part of the corry, formed along with it, and not subsequently by a different operation.

d. The cauldron-like shape of the corry is just what we should expect would be superinduced by an agent formed, moving, and working as a small glacier.

e. These hollows have a decided preference for the higher mountains.

f. They affect the highest parts of mountains.

g. The majority of them face northerly or easterly.

h. According to Professor Ramsay, the corry is eminently characteristic of all glacier countries, past and present.

Notwithstanding this observer's conclusions, an examination for ourselves has led us to believe that ice is only a minor agent in the formation of corrys, they being mainly due to the faulting, jointing, and dislocation of rock masses, combined with marine and glacial action, and probably also with meteoric abrasion.   In some corrys the mass of the drift will be glacial, in others marine, and in a few meteoric; and these different drifts could not have accumulated in such places unless the special forces connected with them respectively had been at work.

In the County Kerry most of the corrys are large and well-developed, while in West Galway and Mayo they range from large to small, Glenawough, in the Formnamore Mountains, being as large and grand as any in Kerry, while some of those situated in the Maumturk Hills are mere holes or recesses in the mountain slopes, some of the smaller being less in extent than the cooms which are now being excavated by the Atlantic Ocean at the present day in the coast cliffs of the neighbourhood.

All the corrys in Galway and Mayo are connected with breaks or dislocations in the strata, and their forms have a greater or less regularity in accordance with the position and number of the lines of breaks. As there is a similarity between the larger excavation (to which some would confine the term corry) and the smaller ones, it is quite legitimate to argue from either to the other; therefore, as the formation of cooses on a sea-coast is easily studied, we may describe their mode of formation.

The north-west coast of the County Galway, near the mouth of Ballynakill Harbour, is favourably situated for the excavation, by marine action, of land-locked cooses, and here, in a short coast-line, many of them may be noted. Of these, one is large enough in plan to be classed amongst Mr Close's corrys, while all are being similarly formed, although many are very diminutive.   This coast is dissimilar to the rest of that of the County Galway, as the cliffs range from 100 to 400 feet high, while it is not open to the full force of the Atlantic Ocean, having nearly a northern aspect, and being subject to a cross current due to the tides meeting the stream flowing into and out of the landlocked bays known by the general name of Ballynakill Harbour; and, as before pointed out, the sea acts more vigorously in confined places than on the open seaboard.

In about half-a-mile of this coast there are seven cooses now being excavated or scooped out; these were carefully examined, and the details noted.

In *Coose No.* 1 there are four breaks in the strata that have facilitated its excavation—a main-break, that seems to be also a fault (*a*, fig. 9, Pl. I.), along which the sea at first worked; it afterwards came to the cross breaks (*b*, *c*, and *d*), when it began to widen out the interior of the coose.

Coose No. 2.—The sea worked in along two breaks until it came to two cross breaks, between which the rock is being excavated.

Coose No. 3.—Here there are two breaks (*a* and *b*, fig. 10, Pl. I.) diverging from the coast-line along which the sea could work, and subsequently excavate out the intervening rock.   The work has now

I

come to a cross break (*c*), and it seems to be going on more rapidly.

Cooses Nos. 4 and 5 respectively are due to two breaks that join at a little distance from the original coast-line.   In both cases the sea at first worked along the breaks, but it is now also cutting out the intervening mass of rock, slight deeps having been formed at the junctions of the joints, evidently in consequence of the broken rock being scooped out by the back curl of the waves round the masses of rock that stand at the entrances to the cooses.

Coose No. 6 is an example of a single break (*a*, fig. 11, Pl. II.) along which the sea is working, and forming a long wedge-shaped gut.

Coose No. 7 is large, well-marked, and extensive, being fully as big as many of the Kerry corrys.   It is about 4000 feet in diameter, a size that is greater than many of the cooms in the Iarconnaught hills. The formation of this coose is clearly dependent on three breaks (*a, b,* and *c,* fig. 12, Pl. II.), two of which met at the old coast-line.   It may be said none of these are large enough to be classed as true corrys.   This we allow, but at the same time it is evident that they are corrys in miniature ; and if the sea had time enough to work, it would enlarge them to any size.

Cooses like that represented in No. 6 are not un- common on coast-lines, while in the hills, cooms of

a similar shape are frequent, on which account they
are passed over without much notice.   Of the other
examples of cooses, if we give due allowance for
subsequent modification by meteoric abrasion, their
counterparts will be found in the cooms among the
Iarconnaught hills. One thing remarkable about these
cooms in the hills is, that although exposed for centuries to meteoric abrasion, their floors are very similar
to those of the cooses now being formed by the sea.
In the hills, west of Lough Inagh, the floors of the
cooms slope outwards ; south of the Erriff valley they
slope to one side, while in the north part of the
Mweelrea mountains they contain deeps ; but the deeps
in the cooms and those in the cooses, as will hereafter
be stated, are not quite similar.   Seemingly the conditions most favourable for the sea excavating a coose
are as follow :—*First*, The rocks on the coast should
lie nearly horizontal or dip inland. *Secondly*, The strata
must be traversed by breaks (whether master-joints
or faults) lying oblique to and crossing one another,
also joining into the coast-line. *Thirdly*, Alternations of rocks of various hardness or composition ; for
if rocks on a coast are homogeneous or nearly so,
marine action will probably denude the coast in more
or less regular sweeps or open curves. And, *fourthly*,
cross currents, in the sea, that will form eddies.   On
an open seaboard, the denuding force must act more
or less uniformly ; not so, however, in partly land-

locked bays or among islands, as in such places
there must be cross currents working in irregularly.
The undermining of a cliff cut by two or more oblique
master-joints or faults may also form a coose, as a
large mass may thereby lose its support and slip out
to sea, to be afterwards broken up and dispersed by the
waves. Such cooses, in different stages of formation,
may be observed on various coast-lines ; but all the
cooms among the hills that may have been formed
from the slipping of a mass of rock are not due to
sea action, as springs, or even the wind, may erode
and carry away some weak beds or strata, and thereby
leave unsupported a mass bounded by master-joints
or faults. In such cases, however, the debris of the
mass will be scattered on lower ground, or occur in
the form of a bar or mound, unless carried away by
ice or a river.

Blanford has pointed out that, in tropical climates,
while marine action is at a minimum, meteoric abrasion is at a maximum ; [1] the quantity of detritus produced by the heat, and carried down by rivers into the
sea, forming large alluvial flats, which protect the
coast-lines from marine denudation ; or in places
where such deposits cannot form, coral reefs grow,
fringing the coast, and having a similar effect.

Explorers of arctic regions have taught us that,
in those latitudes, most of the detritus due to atmos-

[1] "Geology, Abyssinia," p. 151.

pheric influences which is brought down to the sea is
at first left on the " ice-foot," or belt of ice fringing
the coast-line, to be afterwards, when that breaks
up, carried away and deposited in mid-ocean; consequently, in such regions, there can be no natural
breakwater formed, but the coast must be left open
to the full action of the waves and marine currents.
In such places, therefore, marine denudation ought
to be more active than elsewhere.

Let us now consider the temperate zones. Here
all of the denudants do some work; none of them,
however, being at the maximum or minimum of efficiency, as the principal working powers, heat and cold,
under which all act, cannot reach their extreme inten-
sities.   There is, however, some little work done by
each, more or less considerable in accordance with the
varied circumstances under which they act.   Hence it
would seem that *atmospheric influences perform their
maximum of work in tropical climates, and their minimum in arctic regions ;* on the other hand, both *marine
and ice action accomplish their maximum of work in
arctic regions, and their minimum in the tropics.*

Let us now consider the conditions under which
Iarconnaught formerly existed.   Here, as well as over
the whole of the North of Europe, an arctic climate
prevailed at the beginning of the glacial period,
and the efficiency of marine action must have been
considerably greater than at present; to which add

that of glacial action, which does not exist there at
present; and the cooses then formed ought to be much
more extensive, and bounded with higher cliffs, than
those being excavated in the temperate zone at the
present day. We would suggest that the corrys, after
having been in part excavated by marine denudation,
were elevated by the rising of the land out of the
influence of its action, and that afterwards they were
enveloped in snow and ice, and thereby preserved from
being modified by ordinary meteoric abrasion. During the glacial period, all of these corrys, at no matter
what level, were protected by this envelope against
atmospheric influences; but as the climate changed,
those on low ground, especially if exposed to warm
winds or the sun's rays, would be subjected to all the
vicissitudes of the atmosphere, which probably altered
and modified their forms, or even obliterated them;
consequently, while corrys and cooms are numerous
, in high lands, in the adjoining low lands only traces
of them can be detected.

In the northern hemisphere the warm winds and
sun's rays would act more on the south and west
slopes of the hills than elsewhere; consequently, on
the north and east slopes the snow and ice would
remain, preserving the cooms, corrys, and cliffs long
after the south and west sides had been modified and
formed into more or less regular slopes by meteoric
abrasion. That such took place in Ireland is quite

evident, as we before pointed out; for while, as a general rule, that has few exceptions, the cooms, corrys, and cliffs are situated on the east and north sides of mountains, the west and south sides of the hills are slopes. (See page 78.)

It may also be mentioned in favour of marine action having partially excavated the corrys, that such features occur at distinct elevations. Two or more corrys, in the vicinity of one another, may have their floors on different levels, or even in a large corry there may be two or three cooms at different heights; but if the altitude of all the cooms and corrys in a group of mountains or in separate groups of mountains are compared, they are found to be in systems. In Iarconnaught and South-west Mayo, systems occur at heights of 250, 350, 470, 650, 750, 1000, 1200, and 1400 feet; while in Kerry, from the elevations given by Mr Close, they seem to be found at elevations of about 450, 750, 1000, 1250, and 1500 feet in altitude. As, however, there are numerous corrys not mentioned by this observer, this evidence is not conclusive.[1]

[1] All the groups of hills in West Ireland, from County Sligo to County Cork, are, as a rule, much more escarped on the north and east sides than on the south and west. This fact, combined with peculiarities in the drift, led me, on a former occasion, to suggest that, about the end of the glacial period, there must have existed a strong ocean current flowing from the N.N.E. to the S.S.W. ("Notes on Some of the Drift in Ireland," *Dublin Quart. Jour. Science*, vol. vi. pp. 249, *et seq.*), also that to this current the steeps were due. I would now wish to slightly modify this suggestion; for although such a current must have

Subsequently to the corrys having been in part formed by marine action, they seem to have been enlarged and deepened by ice, as ice apparently would have more power than the sea to form the rock-basins so often associated with them; for, as shown previously, it would be more capable of lifting and carrying away all the blocks loosened in the breaks, besides raising additional blocks; for as the water from the ice froze in the crevices and cracks, it would loosen and quarry out other blocks, and raise them up into the influence of the ice-stream.

Although we are inclined to believe that ice was the principal agent in excavating out the rock-basins of the corrys, yet we should point out that water also is capable of excavating somewhat similar hollows, and in an arctic sea may have had powers that are unknown to us of the temperate zone. The formation of those peculiar cylindrical hollows, often called "churn-holes," found in some rocks on the sea-coast, and in river-beds, seems to point to a process by which the sea might grind out deeps in the floors of the corrys. "Churn-holes" are not uncommon, and we find they always occur in rocks containing nodules, concretions, or some other kind of macula

formed cliffs, yet it is probable that they in part existed prior to the current, especially at high altitudes, and were preserved from modification or obliteration by an envelope of snow and ice, while lower down marine action was changing the form of the old ground.

softer or more easily weathered or removed than the rest of the rock. Such spots, when exposed by denudation to atmospheric influence, crack, and subsequently weather out. Once a hollow is formed, no matter how small, particles of sand find their way into it, and these are whirled·round and round, either by wind or the eddying of water, thereby deepening and enlarging it. As the hole increases in size, larger grinders are admitted, and the work goes on more and more rapidly. If the maculæ are situated widely apart, as is often the case, the holes for the most part will be cut symmetrical; but if the weak spots are more numerous, one hole may encroach on another, or, what is not uncommon, holes may form in the side of one in progress, if the grinding process exposes to the weather some of the concealed soft spots. In other cases the maculæ are so numerous that the weathering is quite irregular, and without marked form. Such are generally passed over unnoticed, while the remarkable form of the " churn-holes " always attracts attention.

Similarly, marine action might be able to excavate hollows in cooses. The sea, however, in such cases, could scarcely work as symmetrically; for while, in the case just mentioned, the water would begin its work on the cracks in a more or less regular nodule or the like, the sea in the coose would be working

on a more or less irregular junction of two or more
breaks; but subsequent weathering would modify
the form of the hollow, and make it more bowl-
shaped.

As pointed out by Close, many corry lake-basins
are only in part rock-basins, they being often
partly formed by a bar of drift across the mouth of
the corry, while some only occupy hollows in an
accumulation of drift. In some of the corrys in
which there are rock-basins, also in others in which
there are none, there is a total absence of drift.
This is remarkable, as from C. King's account of the
dying out of the glaciers on the mountains of the
Pacific slope, it would appear that these, the last
parts of the glacier, are enveloped in accumulations
of moraine drift, and we should naturally expect
that in these cold north and east corrys the ice must
have lasted longest.

In some corrys and cooms the lakes are dammed
in by moraine drift, but in others the drift bar is
evidently of meteoric origin. As far as our experience goes, every coom or corry is associated with two
or more breaks crossing one another obliquely; and
if the description of the formation of the cooses on
the Galway coast is referred to, it will appear that
at least one of these break-lines must join into the
old coast-line. This break-line may be perpendicular or oblique to the coast-line; and if the latter,

and there is a sufficient catchment above it, the hollow formed by the break, especially during wet weather, will be occupied by a mountain torrent.[1] These carry down yearly a greater or less quantity of detritus, thereby forming a bar across the mouth of the coom or corry, and impounding the water behind it, thus forming a lake-basin partly rock and partly drift. Such bars are always more or less regular, which distinguishes them from moraine drift similarly placed, as in general the latter is more or less irregular, unless its form has been modified in outline by marine action. Furthermore, meteoric bars are seldom as compact as moraine-drift; consequently, it is not unusual to find lake-basins, so built up, having an overground outlet only in very wet weather, while at all other times the surplus water percolates through the meteoric drift bars.

[1] Mackintosh describes such mountain torrent-courses under the Cumberland name, " Rake."

CHAPTER X.

THE VALLEYS OF SOME OF THE IRISH LAKES.

WE have described the lakes of Iarconnaught, but if we turn from that country to South-west Cork or West Kerry, areas also remarkable for their numerous lakes, we find very similar relations existing between the breaks, faults, and lake-basins ; for although these portions of Ireland are much less faulted than Iarconnaught, yet they are traversed by numerous master-joints and faults, on which the lake-basins are situated. All the areas we have mentioned are more or less mountainous, but the connection between breaks and lake-basins can also be traced in the lowlands of Ireland. Lough Neagh,[1] the premier lake, is evidently situated on lines of breaks, the principal of which bear respectively about N. 10 W., N. 5 E., N. 40 E., and N. 55 E., while there are others of older date bearing more or less obliquely to those mentioned.

Lough Corrib,[2] the second largest sheet of water in Ireland, has also conspicuous lines of breaks,

[1] Admiralty Chart, No. 53, Ireland.   [2] Ibid. No. 2318.

along which lie the different deeps. In a paper written some years ago on this lake-basin,[1] I ignored the open joints, faults, and the like, and tried to prove that the excavation of the lake-basin was due nearly solely to ice action. Since then, however, it has been found that lines of faults or breaks traverse it, while every bay or arm in the N.W. part is connected with one of these lines, and each deep lies along one of them, or at the crossing of two or more; but that ice was the great carrier would still appear correct, as its traces are quite conspicuous everywhere in and about the basin, while the forms of the lake coincide with the courses of the different ice-streams. This lake is partly situated in a carboniferous limestone country, and partly in a country occupied by Silurian and metamorphic rocks; and as the latter rocks are much more broken up and faulted than the former, the most of the part situated in the limestone, is much shallower than the rest. There are two places in which there are deep abrupt holes, that could not have been scooped out by ice, and probably were formed by water that had a subterranean passage, assisted by a fissure due to a fault or shrinkage. These deeps will again be mentioned when we are describing the Lough Mask basin, but prior to entering into that subject we should say a few words on *Sluggys*, *Swallow-holes*, *Turloughs*,

---

[1] *Geological Magazine*, Nov. 1866.

Pot-holes, *Pipes*, and the like, as all the low-lying lakes in Ireland seem to have now, or to have had at a former period, some connection with them.

We have already pointed out how a portion of land may sink if an underlying subordinate bed or portion of a bed is subtracted by water. Sometimes these subsidences form bowl-shaped hollows, but at other times a regular hole will form, called a sluggy or swallow-hole in Ireland, and a pot-hole or pipe in other places. Sluggys generally are somewhat like the shape of an hour-glass or of a funnel; they may, however, have vertical sides. Pipes and pot-holes in limestone and chalk countries are supposed by sub-aerialists to be "produced by the slow dissolving action of water charged with carbonic acid." This view seems to be now generally adopted; but although in some few cases carbonic acid may act alone, yet it appears to us to act in most cases in conjunction with lines of joints, structure or other weaknesses in the rocks. It is argued that because some small pipes can be proved to have been excavated by dissolving action, that all large pot-holes are similarly produced; this, however, seems to be a weak argument. In various places in Ireland holes similar to what are called "pipes" in England are in course of formation among the tracts of carboniferous limestone, and from these we learn that such holes may be formed by two different processes—one variety being excavated from the top downwards, and the others mostly from the bottom upwards. The growth of the first may be studied on the shores of any lake situated in a limestone country such as Lough Corrib. Here, when the drift covering has been recently removed from the rocks by the water of the lake, beautifully polished, etched, and scratched surfaces are exposed. Those that now, since the lake was lowered,[1] are exposed only during the summer months, are full of minute round holes; those that were formerly exposed only during the summer months are full of larger holes; while those that are exposed to all weathers are pierced through and through with circular holes. This circular weathering is peculiar, as it does not seem to occur except in the vicinity of lakes. It usually begins on the ice-polished surfaces by forming lines of minute holes along the striæ; but if the rock is affected by joints, the holes will more rapidly increase along the joint lines, so that in a very short time all appearance of the striæ disappears. Moreover, even if the rock is not jointy, all appearance of the striæ will shortly disappear, as these scratches run regularly alike over the chemically hard and soft portions of the rock, while the latter will weather much more freely than the former. Professor Melville of Galway seems to be of opinion

---

[1] About twenty years ago, during the works for the Lough Corrib Navigation, the summer level of the lake was lowered three feet.

"that these holes are caused by the acid generated from the decay of mosses or lichens, humic acid being formed from the decay of the vegetable matter that might grow in or be swept into the small hollows in the stone, and carbonic acid from the final decay, the latter acid saturating the water, which would then act on the limestone. Once the hollow is begun, all acid in the water remaining on the stones, along with what may form from the decay of any vegetable matter that may be swept into them, will be concentrated as the water evaporates; therefore principally act on the bottom of the hollow, and thereby bore the cylinders through the blocks."

In this way the holes which are like inverted funnels might also be explained, for as such holes deepened, the water could not, in them, evaporate as quickly as when they were shallow; therefore the contained acid would act on the sides as well as on the bottom of the holes, and thereby enlarge as well as deepen them. If the acid generated from the plants is the cause for these weatherings, the reason for their being only found in the vicinity of lakes may be accounted for by supposing that the moisture from the lakes helps the growth of the acid-forming vegetables.

Against these suggestions it should be mentioned that in one locality, Dringeen Oughter, on the shores of Lough Mask, a cup weathering occurs, not on the surface of the rocks, but on the surface of one of the beds in the face of a low perpendicular cliff. This cliff margins a turlough, and the usual winter level of its water seems to agree with this line of weathering, the note made being, " Cup weathering at the top of the winter floods, but over it are two feet of rock; these holes may be possibly due to the decay of plants."[1]

Away from lakes a weathering generally takes place in connection with the joint-lines. If the rocks are thin bedded, it breaks them up into a very coarse shingle, but if thick bedded, the weather will open narrow lines of more or less deep fissures. Such fissures, although connected with, are not the true pot-holes, as for the latter to be formed there must be portions of the rock of a softer consistency than the rest, so that some places in the mass will weather more freely than the rest, in which places the pot-holes form. If due solely to the dissolving action of acid on a homogeneous rock mass, pot-holes ought to be circular. This, however, is rarely the case, as in general they are more or less rudely oblong; their longest axis always coincides with the jointing of the rocks. Further more, the pot-holes usually occur in systems of lines; and our colleague, Mr W. Topley, F.G.S., when writing of the pot-holes in the chalk bounding the Medway valley, says, " The marked parallelism of the long pipes at Maidstone is an argu-

[1] " Memoirs of the Geol. Survey of Ireland," Ex. sheet 95, p. 47.

K

ment in favour of their having been originally started along joints or fissures."

A pot-hole formed in a sheet of limestone will have a deposit of clay at the bottom of it, the insoluble residue left after the rest of the rock has been carried away in solution through the cracks and joints of the rocks, while if the rock contains nodules of chert their debris will be mixed with the clay. It is the same with pot-holes under drift, except that not only the bottoms, but also the sides, will have a clay lining. This is as might naturally be expected. As the holes became deeper and wider, the drift coming in from above would force out the clay and form a coating round the walls ; or if the holes were filled by a more recent accumulation of drift, during the subsequent weathering the water percolating down over the bed of limestone would carry the residue along with it, and thus form the clay lining. It has been suggested that the clay in the pot-holes in the chalk came in from above. This may possibly be the case in some instances, but in most it is impossible, as the clay at the bottom of the holes and lining the sides usually far exceeds in quantity the clay that would be necessary to form a horizontal sheet over the top of the hole, if of an equal thickness to the layer of clay that all round the pot-hole intervenes between the drift and the subjacent rock. In some of the subterranean caves brought to light during quarrying or other works, thick deposits of clay occur, evidently the insoluble residue of the limestone. As this clay is exactly similar to the clay lining the pot-holes, it seems to suggest that both were due to the same cause.

The swallow-holes, formed from below, mark the lines of subterranean streams. In company with our colleague, the late Mr F. J. Foot, M.A., &c., we explored many of these passages, they being of frequent occurrence in the Counties Clare and Galway, and in all places found that they were induced by joints and other breaks in the strata along which the water worked, gradually widening them, partly by dissolving away the rock; but after the passage had become large, and the stream had increased in size from the surface drainage being diverted, the current acted more like an overground one, abrading as well as dissolving away the rocks. In underground rivers and streams during flood, when the passages are choked with water, the latter acts upwards in jointed parts and other weak portions of the roof, till eventually an opening connected with the surface is formed, in consequence of a portion of the superincumbent mass giving way and falling in, while afterwards the holes are gradually increased in size by meteoric abrasion. Pot or swallow holes in all stages of progress may be seen in the plains of Galway and Mayo; the rock sometimes being covered by boulder-clay or esker-drift, at other times by a meteoric accumulation formed of the insoluble portion of the limestones that has weathered away. If a swallow-hole forms in a low place, the river during floods will overflow, and form a lake locally called a "blind lough" or turlough. These usually exist in wet weather only, but totally disappear afterward, leaving in their place a rich meadow or pasture. Of one of the turloughs in the County Galway there is a tradition that it was formed by artificial means, a well having been sunk that tapped a river and flooded the hollow, occupying it ever since. In Galway and Clare extensive areas and lakes are drained through subterranean rivers, the outlets from some being unknown.[1] This occurs in other places also in Ireland; while it appears probable that in ancient times, when the land was much higher than at present, some lakes which now have surface outlets previously had subterranean ones. A description of one of these river systems, devoid of an open outlet to the sea, may be given.

Lough Cooter receives the drainage of the N.W. shoulder of Slieve Aughta. The Beagh river leaves the north end of Lough Cooter, and flows for about two miles towards the west, when it disappears in a

---

[1] "Memoirs of the Geological Survey," Exs. and Maps, sheets 115, 116, 124, &c.

cavern. From this its course can be traced for about six furlongs by swallow-holes, called the *Devil's Punch-Bowl*, the *Blackweir*, the *Ladle*, and the *Churn*, to *Pollduagh*, a cave out of which it rushes again to daylight. From this, under the names of the Cunnahowna and Gort river, it runs by Gort in a northerly direction for about three miles, when it sinks a second time, about half a mile S.E. of Kiltartan, at the old castle. From this it runs under ground for about six furlongs, and comes to the surface immediately west of the village of Kiltartan. After sinking and rising two or three times in Coole demesne, it eventually rises S.E. of Raheen House, and from that flows into Coole Lough.

The Owenshree river, which flows by Roxborough, takes the ground half a mile W.N.W. of Castle Daly Roman Catholic Chapel, and rises again at Coy Lough, about two miles N.E. of Kiltartan, where it joins into the Boleyneendorrish river. This takes the ground in less than half a mile, but eventually appears near Raheen House, and having mingled with the Gort river, flows into Coole Lough. From Coole Lough the waters find their way in subterranean passages by Caherglassaun Lough to the sea at Kinvarra, where part, at least, finds egress through the joints of the rocks in the vicinity of Dungorey Castle.

In Coole demesne a subterranean stream flows

southward from Corker House to join the open river. This must come from the Boleyneendorrish river, as when there is a freshet in that river, and none in the Gort river, the river at Coole becomes quite muddy; but if there is a flood in the Gort river, a stream flows from Coole up this subterranean passage towards Corker House. Where the Gort river takes the ground, on the S.E. of Kiltartan, there are the traces of two older channels at higher levels than that through which the water during the dry season now finds its exit. The highest level is now entirely stopped, and a stream flows back along the old course down to the present river. The middle level, during floods, is still used, and the water rushing down into the subterranean channel below forms a small whirlpool.

A very large turlough is formed in connection with Coole Lough. During the dry weather the water on every side flows into this lough; but when the floods arise, the subterranean passages are not large enough to carry off all the water, which therefore fills the lough and overflows to the south, where it forms a most extensive turlough in the neighbourhood of the Newcastle racecourse, which lies to the N.W. of Gort.

On the 12th October 1862, cricket was played on the racecourse of Newcastle, and there was scarcely any water in the streams, but the succeeding week was very wet, and on the 21st the whole of the plain was covered with water, over twenty feet deep in places. The area under water, including the flooded land about Coole Lough, was at least 500 acres; but when we consider the extent of the water-basin which drains into Coole Lough, it is not surprising.

Some of the lakes and turloughs on the low ground are affected by the rise and fall of the tide, the rising tide damming up the egress of the fresh water, which accordingly rises in the lakes or turloughs. This cannot be observed during floods, as from the expanse of water the rise and fall would only be a few inches, but during dry weather it is most perceptible. This phenomenon can be very well seen in the holes in the vicinity of the mine on the west of Caherglassaun Lough, although six or seven miles from the sea at Kinvarra.

Lough Mask lies on the north of Lough Corrib, and is in some respects similarly circumstanced, as the rocks of the area occupied by its basin are in part of carboniferous age and partly much older, the latter being greatly broken up by faults and the like. The lake-basin[1] has a bearing of about N. 20 E., similar to the general bearing of the main joints in the country on the east, while to the S.W. are two arms or bays, branching from the main basin. The greatest

[1] Admiralty Chart, No. 2318, Ireland.

deep of the lake lies near the western shore, and has
the same bearing as the lake itself, the main joints,
and also the ice-striæ; from which it would appear
that ice, taking advantage of the joint-lines, was the
main excavator of the basin, more especially as all the
deeps, except in one place, are more or less gradual
and regular in outline, and might have been cut by
an agent going from the S.S.W. toward the N.N.E.
In this deep place there are no data by which a connection between it and the breaks in the country on
the west can be traced; but we know that the country
to the N.W. of it is very much cut up with faults,
while the rocks are jumped and shifted; one large
fault, a downthrow to the N.E., striking for the
S.W. end of the great deep (over 150 feet), the deep
having a transverse widening in its line of bearing.
The bays forming the S.W. arms of Lough Mask
are each known to occur along a line of fault, while
we learn from the chart that the different shallows
and deeps occur respectively on the up- and downthrow sides of the faults, the changes from a shallow
to a deep hole being quite sudden. Lough Mask
drains by subterranean passages through the barrier
of limestone that on the south-east divides it from
Lough Corrib, and in the deep just mentioned there
is a hole remarkable for being on nearly the same
level as the deepest hole in Lough Corrib, the former
being 127 feet lower than the Ordnance datum line,

and the latter 124 feet, which would seem to suggest
that these holes may have been the connection
between the subterranean passage and the drainage
of the valleys when the land was at a higher level than
at present. When this basin, and also that of Lough
Corrib, were filled with ice, the water associated with
the latter must have found a vent somewhere, and
probably it was through such deep places; while after
the ice disappeared these may also have acted until the
land sank so low that there was no fall from them, or
until they were silted up by matter carried into them
by rain and rivers. Lakes in a limestone country are
usually very irregular in outline, and at first it might
be supposed that there is no connection between their
shape and the structure of the subjacent rocks. On
an examination, however, it generally appears evident
that the bays and all wide stretches across the lake-
basins conform with lines of breaks or displacements,
while the minor features of the coast-lines are due to
the weathering along joint systems or lines of bedding, generally the former.

The basin of Lough Derg, through which the
Shannon flows, well illustrates the effect of breaks on
a lake-basin. On looking at the chart of this lake,[1]
it will be seen that by a few displacements the basin
might be made to occupy a N.N.E. and S.S.W.
valley, and that each of these displacements nearly

[1] Admiralty Chart, No. 1552, Lough Derg, Ireland.

coincides with the strike of the direction of the pre-
viously-mentioned faults connected with the groups
of hills called Slieve Bernagh and Slieve Aughta.
From Killaloe to Rinnaman Point, the bearing of
the lake is nearly south and north; there the direction
changes, although no fault is known to cross the basin.
From Rinnaman the deeps extend for four miles
towards the N.N.E., when they are probably crossed
by a fault, north of which they bear north and south;
this fault, however, has not been proved, but that it
exists seems probable, on account of the position of
the old red sandstone in the hill N.E. of the Cornagoe
valley.

A little farther north the deeps run into the great
east and west reach that extends between Scarriff and
Youghal bays. This east and west stretch lies in
the continuation of the limestone valley which
separates Slieve Bernagh from Slieve Aughta, and
this limestone seems to be brought into its present
position, not only by a synclinal curve, but also by
faults. On the north of the valley a large fault, a
downthrow to the southward, has been proved, while
near the south margin one or two lines of breaks
seem to exist; but they could not be proved through
the valley on account of a great accumulation of drift
obscuring the geology hereabout. One of them was
proved, however, farther west in the low country in
the line of this valley, and in its strike are all the

deeps that occur in the great east and west expanse of
Lough Derg. At this reach the S.S.W. and N.N.E.
lake valley seems to be shifted eastward nearly three
miles [1] to the entrance of Youghal Bay, from which
the deeps extend N.N.E. to a little south of Illaunmore, where they are crossed by the north fault of
the Scarriff valley, which changes their bearing to
nearly north and south, while farther N.E. of Coose
Bay, the great deeps (over 70 feet) end where the
basin is crossed by the eastern continuation of the
Cloonnagro and Corra valley fault. In the north-
eastward of Coose Bay, the trough of the basin runs
nearly N.E. and S.W. to off Drominagh Point, near
which it crosses the fault supposed to run nearly
west and east from Lough Atorick; while farther
north, an east and west reach, extending between
Cloondavaun and Terryglass bays, lies on the eastern
extension of the fault proved at Woodford. Still
farther northward, the original lake-basin extended
in a N.E. direction, and the relations between it and
the other faults in Slieve Aughta could be pointed
out; but as this part is now silted up, nothing further
need be said about it.

There are also other breaks that appear to come
into the Lough Derg basin from the S.E., as a rude
parallelism exists between the bearings of the

[1] Such a shift may appear enormous, but as large a shift has been
proved in the Maum valley, County Galway.

different bays. We have not, however, as carefully examined any of this ground, except the valley of Youghal Bay, in which a fault evidently exists. This lake is one of the natural reservoirs on the Shannon, and at certain seasons the water is very muddy, while at other times there is a great evaporation from it, which causes large deposits of mud and marl; therefore from the soundings very little can be learned as to the nature of the original bottom of the basin, in addition to the two facts of the deeps being in the lines of the break, or at the junction of two or more; and of the islands, rocks, and shoals being on the upthrow sides of the faults. This latter fact is very apparent, except in the great east and west reach, on account of the throws of the south faults of the Scarriff valley being undecided.

We have next to consider how the rock-basin of Lough Derg was excavated, for although the water is now, in places, edged with drift or alluvium, yet formerly the area was nearly, if not quite, surrounded by rocks. The sea probably was the first excavator, marking out, along the faults and breaks, valleys and ravines afterward to be occupied by ice and rivers, but when it retired, the other denudants did more or less work. Principal Dawson, LL.D., M'Gill's College, Montreal, has pointed out, in relation to some of the large North American lakes, that apparently the original surface passages from them

have been obliterated, and the water forced to find
other outlets. This may also have been the case
in regard to the basin of Lough Derg, as the valley
extending from the long east and west reach through
Youghal Bay round the hills called Slieve Arra, and
down the valley of the Kilmastulla river to the
valley of the Shannon, a few miles south of Killaloe,
is occupied by an accumulation of deep drift, principally gravels and sands, apparently of the " Esker
sea " period. This drift may be of any depth, and
fill a deep valley; and if such a valley once existed,
meteoric abrasion, or even sea action, would have
been capable of excavating out the basin of Lough
Derg about 60 feet below the present surface
of the waters of the lake, that is, down to a level
nearly equal to the height of the barrier of rocks that
form the rapids and fall of Doonass in the neighbourhood of Castleconnel. Such a passage would drain
all the upper part of the lake-basin, except four
small deeps, one northward of Illaunmore, one immediately west of that island, and two to the S.S.W.
of it; but it would be incapable of draining a long
narrow tract in the S.W. arm of the lake, also a
portion of the east and west reach, they respectively
being over 60 feet in depth, and in many places
over 100 feet; the deepest place, which is situated in
the latter, and lies a little to the N.E. of Parker's
Point, being 119 feet. Mr John Ball, F.R.S., &c.,

in a paper on " Soundings in the Lake of Como," [1] has pointed out as to a lake-basin, " Supposing the rocks on either side to be of equal hardness and similarly stratified, it is safe to affirm, that if they had been hollowed out by glacial action or by aqueous erosion, the slope would be steepest on the concave side of the bend, in those parts of the lake where the glacial stream was turned aside from its previous direction." This Ball has shown is not the case in the basin of the Lake of Como; neither is it the case in the Lough Derg basin, as at all such bends the slopes are steepest at the convex side of the bends, while the deepest part of each cross section is nearest to the same side.

After carefully contouring and examining the charts of three Irish lakes, Loughs Corrib, Mask, and Derg, we have found, in all, sudden deep holes which it seems impossible either the sea, ice, or meteoric abrasion could have excavated; while as they occur on or at the junction or crossing of breaks, we would suggest that they are in part due to the fissures which were formed by the contraction of the rocks, and that subsequently most, if not all of them, may have been connected with subterranean passages, which at different periods drained or helped to drain the lake-basins. If the Lough Derg basin had been, for ages, occupied by a

[1] *Geological Magazine*, vol. viii. No. 8, August 1871.

glacier, most, if not all the loose portions of the rocks should have been removed out of it, leaving the ice, prior to its final disappearance, comparatively speaking free from rock detritus contained in it, to encumber the lake-basin. At the present, the deepest spot in the lake is only eleven feet below the low-water of ordinary spring tides at Limerick, the soundings on the chart having been reduced to the level of 108 feet above that height; but as the level of the water of the lake is so much higher, there would be an underground drainage if these holes were connected with subterranean passages. During the glacial period, the water under the ice would have found vent through these passages, denuding away their walls, and enlarging the deeps from which it flowed. After the glacial period, water may still have flowed through these holes, until eventually the land sank so low that they were incapable of acting, or they were choked by the deposition of mud or marl. But even now there is reason for supposing that such passages may be partially open, as in the summer less water leaves the lake-basin at Killaloe than flows into it by the main feeder at Portumna and from the numerous smaller tributaries. Part of this deficit is certainly due to evaporation, but all of it can not thus be accounted for.

Many of the Irish lake-basins which now have

surface streams flowing from them, especially when in the carboniferous limestone, may have had an underground drainage. This is suggested by the many systems of lakes connected with subterranean rivers that still exist in different parts of the island. We have already described those connected with Coole Lough, near Gort, County Galway, and we may now give a brief description of the river Fergus, County Clare, from the pen of the late Mr F. J. Foot, more especially as the lakes connected with it are of considerable size.

" The river Fergus rises in Lough Fergus, between Corrofin and Ennistimon, at an elevation of about 350 feet above the sea ; flows eastward and northward for two miles and a half, when it receives the Clooneen river ; thence it takes an easterly course for a mile and a half, when, on entering the limestone ground, it suddenly disappears in a swallow-hole or vertical cavity in the rock. Half a mile to the east it again emerges to the light, from a cavern called ' Poulnaboe,' from which it flows down into Inchiquin Lough, and thence into Lough Atedaun. No visible river flows out of this lough, but the Fergus is supposed to have a subterraneous course in a direction of about E.S.E. to Dromore Lough, whence it flows southward, now above and now below ground, to Ennis, and thence to the Shannon."[1]

[1] " Maps and Memoirs, Geological Survey," Sheets 122 and 123.

If the deeps in the basin of Lough Derg below the 50-feet line of soundings were mapped, they would make a system of lakes somewhat similar to those of this part of Clare, provided that each deep were connected with the next by a subterranean passage, while the great deep in the east and west reach had a passage opening somewhere below the present sea-level.

Before leaving these lakes, we may refer again to Lough Corrib, as the river flowing from it has suddenly become dry four times during the historical period, namely, in the years A.D. 1178, 1190, 1647, and 1683. In the paper previously referred to,[1] I suggested that these failures of water might be accounted for by subterranean passages, as during very low tides the water from the lake would force out the sand, or whatever else choked these up, while the return of the tide would fill them up again; and if such passages do exist, they must be connected with the former drainage of the basin.

[1] Page 141.

CHAPTER XI.

GORGES AND RAVINES.

In Great Britain and Ireland most of the valleys
are more or less wide, with sloping sides, and have
evidently suffered from various denudants since they
were first formed, having been probably two or
three times submerged and exposed to marine
denudation, and as many times raised up, so as to
come within the influence of the atmospheric agencies
or ice. In tropical climates this does not seem to be
so often the case, as many gorges and ravines are
well developed, and apparently the only denudant
that has acted on them is meteoric abrasion. To
those in Abyssinia we may draw attention, as they
have been lately described by Mr W. J. Blanford.[1]
This observer has stated that no faults seem to exist
in the ravines, and that they could not possibly have
been excavated by either the sea or ice action; but as
he does not take into consideration the contraction of
rocks as they come to the surface of the earth, he is
led to believe the valleys were formed solely by
meteoric abrasion. Such a theory would require an

[1] "Geology of Abyssinia," pp. 86 and 87, 154 and 155.

enormous length of time for their formation; while
in one instance, that of Lough Ashangi and its
tributary streams, the author cannot account for the
disappearance of the detritus excavated out of the
tributary valleys. If, however, it is allowed that
rocks, when brought up to the surface of the earth,
contract more or less, and form open fissures—not
necessarily vertical displacements, but only breaks in
the continuity of the beds—such gorges as those in
Abyssinia would be easily accounted for, without
either a long period of time being expended in their
formation, or any large quantity of matter having been
denuded out of them.

In such a narrow valley as the fissure formed by
the contraction of a mass of rocks, denudation from
water would principally take place at the bottom,
deepening it more quickly than it would widen it,
except in a very damp climate, where a protecting
envelope rapidly grows which might prevent the
bottom of the ravine from being denuded; this may
be due to different circumstances. If trees or other
protecting plants grew on the tops of the cliffs, all
meteoric abrasion would be prevented from working
downwards; consequently it could only act horizon-
tally, or inward from the sides of the ravine. Or a
hard rock may cap soft ones—soft sandstones in
Abyssinia being under bedded igneous rocks—this
also would prevent denudation from above. Besides,

as previously stated, inland perpendicular cliffs,
especially in friable rocks, are inclined to remain
perpendicular even in our temperate climate, as
the denudants that principally act on them are the
wind and heat; and if wind can act so effectively in
this country, what would it not be able to perform
in those countries where the heat is excessive and
hurricanes common.

It has previously been pointed out that some of
the ravines in Iarconnaught seem due nearly solely
to the contraction of the rocks, and greater effects
might be expected in a country like Abyssinia,
where the rocks are exposed to a tropical heat; and
if at the first these deep narrow gorges were thus
opened, rain and rivers would deepen them, while
the nearly perpendicular character of their sides
would be preserved principally by the action of the
wind, denudation from above being prevented by
the luxuriant growth of tropical vegetation, or by
the capping of igneous rocks.   Of debris to be
carried away by the streams there ought to be very
little in addition to that abraded off the bottom of
the gorges, as nearly all of what was denuded by the
winds would be borne away upwards to be scattered
over the high lands.

The contraction of the rocks would account not
only for the formation of the ravines, but also for
lake-basins like that of Lake Ashangi, without the

aid of much denudation or any convulsion of Nature.
It has been already shown that when two or more
fissures meet or cross each other, a greater or less
vacancy is formed (fig. 5, Pl. I.); and from an examination of the geological map accompanying Blanford's
description, we learn that various valleys radiate from
this lake-basin, and if valleys, probably also breaks.
Here, therefore, a greater or less vacancy ought to
have been formed, and after it had been modified
by meteoric abrasion, it would be represented by the
present lake-basin. Blanford has suggested that
this lake-basin " must be of very small geological
antiquity, or it would have filled up long since." This,
however, would be quite unnecessary if the valleys
and basin are principally due to the shrinkage of
the rock; as then the materials to be denuded would
be of but small amount, and any that did go down
may be yearly carried away by a subterranean passage
similar to that which Blanford supposes to exist;[1] as
other lakes with similar outlets have been found to
be capable of discharging the detritus brought into
them by streams. This is particularly evident, both
in the Counties Clare and Galway; as many of the
lake-basins, there, would long since have been filled
up by the solid matter borne into them, were it not

---

[1] Mr Blanford points out that circumstances over which he had no
control prevented him giving the time he would have wished to the
examination of the Abyssinian lake-basins, but to this lake-basin he paid
particular attention.

afterwards carried out of them by the water. The power of water to keep clear a river-bed having an underground vent, is exemplified in the Beagh River, County Galway, the stream that flows from the north end of Lough Cooter. This river, after an open course of about two miles, becomes subterranean. At and near its disappearance its bed is enclosed by high banks of boulder-clay drift, which contain blocks of limestone and sandstone. These banks are yearly more or less denuded, and the blocks in the drift are numerous; yet after a flood few of them remain in the bed of the river, it being for the most part occupied by a flinty, angular gravel lying on the denuded limestone rock. During low water no open vent seems to exist through which the blocks could be carried, the water flowing into a pool with a muddy bottom, at the base of the cliff. The limestone blocks might be in part dissolved and carried away in solution: this, however, could not happen with the sandstone. But a close examination explains the process. The vent, as viewed during low water, is a filter of stones, with a greater or less coat of slime; when, however, a flood comes down, the mud and stones are disturbed and a passage opened, which increases more and more till a whirlpool is formed by the rush of the water that rolls the stones over one another, and thus grinds them up. Some may even find their way into the subterranean passage; as lower down, where the stream again comes to daylight, blocks and fragments of sandstone have been observed.

The previously described lakes of Clare, which have subterranean outlets, act somewhat similarly; for if rain comes on after a drought, they rise at first rapidly, while subsequently, although more water may be flowing into them, their surface will be lower, showing that at the first the passage was stopped by a deposition of mud or marl, which the rush of water had first to dislodge, and then carry away, before obtaining a free passage for itself to run off.

Similar reasoning may be applied to account for the present condition of Lake Ashangi, as all detritus washed into the lake may be disturbed and carried away during sudden floods. It may be objected, that if detritus can be thus carried away through subterranean passages, it is quite unnecessary that the gorges and lake-basins of Abyssinia should be supposed to be due to surface shrinkage, as all the detritus denuded out of them may have been thus disposed of. But let us not slight the significant facts that the basin lies at the junction of a number of valleys; that as "no ice ever existed in the country,"[1] there was no denudant capable of excavating out the lake-basin; that as "there is a subterranean passage from the lake,"[2] there must be a fault or break connected with its basin; and that

[1] Blanford. [2] Ibid.

prior to the formation of the lake-basin there was no
receptacle for the deposition of the debris denuded
out of the valleys ; even if we quite ignore that one
process would take ages for its accomplishment, while
the other would require a comparatively short space
of time.

In one of the "Memoirs of the Geological Survey
of India," Mr A. B. Wynne, F.G.S., gives a map and
description of Mount Sirban, in the Upper Punjaub.
The hill is thus described :—" It has an elevation of
6243 feet, with an elongated oval base, is entirely
isolated from the surrounding hills by deep narrow
glens, and is penetrated by steep-sided ravines,
radiating from the summit. Its form is massive,
and its outline heavy, presenting convex curves to
the northward, suddenly interrupted by precipices,
which, facing to the southward, extend along its
crest, and send rugged broken spurs into the valley
below." Wynne seems to have been principally
employed in determining the age of the different
rocks entering into the structure of the mountains,
yet he ascertained that the principal " passes or
valleys " coincided with lines of fault, and in his
sketch-sections the relations between the features of
the hills and the lines of breaks are most evident.
Also, Mr J. C. Ward, F.G.S., an extreme subaerialist,
has to allow when describing the scenery of the Lake
District, England, that " lines of faults are often

found to run through its mountains-gaps or passes."

The gorges that have been mentioned seem principally due to faults and shrinkage fissures, but there are others that appear to be nearly solely due to the weathering out of courses or dykes of rock. Those due to soft or "rotten" courses of rock often occur in granite countries, where narrow, deep passes, with steep or overhanging walls, often cut through a hill or mountain range. Why these narrow, rotten courses should occur in granite has not as yet been explained; but they weather much more quickly than the mass of the rock; this being probably due not only to a different chemical composition, but also to their being intersected with innumerable small, irregular shrinkage fissures or cracks, through which frost, heat, rain, and wind easily penetrate into the rock, disintegrating it, and carrying away the particles, thereby quickly forming more or less deep, narrow ravines. Small ravines of this class are not uncommon in our granite-hills, but they are seldom as well developed or as striking as those of some other countries. Mr John Muir, when describing the newly-discovered "living glaciers" in California, mentions the "narrow-slotted cañons, called devil's lanes, devil's gateway," &c., which are often from 100 to 200 or more feet deep, while they may be only 10 or 15 feet wide. Many of these gorges he finds

to be filled by glaciers " still engaged in cutting into
the mountains like endless saws."

Some such ravines may possibly be due in part to
shrinkage, as it is not uncommon to find between
the rotten granite and its walls a thick parting of
flucany stuff, or a thin rib, or even a good metalliferous vein; any of which would be due to a shrinkage fissure. These partings and ribs have been observed along the walls of many of the courses of
" rotten granite " in West Galway, while some of
the metalliferous veins in Cornwall are connected
with courses of " soft growan ; " and from Wyley's
description of the mines in Namaqualand, South
Africa, it would appear that some of those lodes are
similarly circumstanced.

The denuding out of dykes of fault-rock, and
consequent formation of gorges or narrow fissures,
have already been described, and we may now add
that dykes of igneous rocks, when weathered away,
may leave very similar passes. These, indeed, may
have been at first shrinkage fissures, which subsequently were filled with the intrusive rocks; as it
is not uncommon to find the rock on both sides of
a dyke unbroken or unbent, which could scarcely be
the case if the igneous rocks had forced open the
fissure they now occupy; besides, in some cases ribs,
and even thick mineral veins, occur alongside dykes
of igneous rocks, as if there had been a shrinkage

after the first filling of the fissure, which has caused
a second fissure, now occupied by the minerals. To
return, however, to the gorges now in question—no
matter what was the original cause of the fissures,
the present gorges are due to the weathering away of
igneous dykes, which happen to be in these cases
more easily denuded than the associated rocks. In
the west of Galway and Mayo there are some
marked passes of this kind which for miles have
nearly perfectly level floors; these are locally called
" Bohernacolley " (*The hags' road or path*), as they
are supposed to have been formed by witches.
Here the dykes are generally a variety of whinstone
which weathers rapidly, principally by chemical
action, leaving standing on both sides walls of granite
or metamorphic rocks. In other places, however,
there are dykes of felstone, which at first sight might
be expected to resist denudation better than the
associated rocks ; but they do not, on account of innumerable systems of joints, which break them up
into a small angular gravel. Here it should also be
mentioned, although not connected with gorges, that
some whinstone dykes in that country, although
to the eye very similar to those that weather away,
have resisted denudation better than the metamorphic
and other associated rocks, and now stand out in
thick wall-like masses, stretching in more or less
regular lines along and over the hills.

CHAPTER XII.

THE RIVER VALLEYS OF SOUTH-WEST IRELAND : THE VALLEY OF THE WEALD, AND THE FIORD OF KILLARY HARBOUR.

In a paper "On the Mode of Formation of some of the River Valleys in the South of Ireland,"[1] the late J. Beete Jukes, F.R.S., suggested that subærial denudation was the principal carver of the physical features of the south of Ireland. The reasons on which this eminent geologist founded this theory may be epitomised as follows :—1st, Limestone once existed over the whole of the S.W. of Ireland. 2d, The valleys are not connected with faults or fissures. 3d, Marine denudation only acts with a broad, horizontal movement, tending to plane down the land to its own level. 4th, Marine denudation cannot produce ravines or narrow, winding valleys. 5th, All glens, ravines, and winding valleys have been excavated by either ice or rain and rivers; and 6th, That the internal forces of disturbance must have ceased to act long before the present surface was formed.

[1] *Quarterly Jour. of Geol. Society of London,* Nov. 1862, p. 379.

The first of these propositions Mr Jukes subsequently gave up, and had to allow that the opinion of more than one of his colleagues was probably right, namely, that the carboniferous slate and Coomhola-grits of S.W. Cork were the representatives of the carboniferous limestone of the central plain, and that the latter rock never existed in that country. This, however, does not much affect the present subject, as some of the other rocks are nearly as easily denuded as limestone. Of the other propositions, none seem to us to be proved. We shall begin with the second and sixth taken together.

In the river valleys of S.W. Cork we find the geology concealed by alluvial deposits, the relations of the rocks on one side of these flats to those on the other being only conjectural, and on account of the high dips of the strata, considerable faults might exist without their being observed; which seems all the more probable, from the fact of rocks sometimes coming down one side of a valley while there are none on the other side.[1] Besides, all the valleys in the country that extend northward and southward are in

---

[1] In the valley of the Boyne, where the viaduct crosses the river, in the vicinity of the town of Drogheda, there is no surface appearance of a fault further than the river valley; yet when sinking the foundation for the piers of the bridge, at 20 feet on the Meath side nearly horizontal calp limestones were found, while for the Louth pier an excavation over 90 feet deep was sunk without coming to any rock, there being between the two places a great downthrow to the northward.

more or less parallel systems, which includes not only
the valleys, but all associated faults and master-
joints. In the river valleys the connection between
the valleys and faults may not be apparent; but on
an examination of the cliffs, it is found that there is
not one of the fissures extending to the coast-line
which is not connected with a break or fault in
the underlying rocks. Furthermore, in those portions of S.W. Cork where mining operations have
been carried on, slides, heaves, or cross-courses have
been proved under every transverse valley, ravine,
fissure, and river or stream course that has been
mined under; and such facts suggest that all the
transverse river valleys in S.W. Cork are connected
with lines of breaks or faults. The longitudinal
valleys have still to be considered; but as no mining
operations have been carried on across them, and as
they are more or less filled with boulder-clay or moraine drift, an examination of them is very unsatis-
factory. Although faults can only be proved to extend
along one of them, yet it is most probable that they
do exist in all. Jukes states that it is probable that
a band of limestone extends up the valley from Killarney to Mallow; this, however, is far from being
proved, as S.W. of Kanturk the coal-bearing beds of
the coal measures are found close to the Old Red
Sandstone, while from that westward to Mill Street
there is scarcely room for any limestone to come in

between the coal measures and the Old Red Sandstone. Even if there are limestones, it is scarcely possible to deny that a great fault, with a downthrow to the northward, exists here, which can be traced from Dingle Bay on the west, past Mallow towards Mitchelstown, while a branch (pointed out in chap. vii. p. 101) seems to run to Dungarvan. In this line of country are situated the valley of the Glenflesk River and the major part of the valley of the Blackwater; consequently neither of these valleys can be said to be unconnected with faults. To the south of this line of fault all the rocks are twisted and folded, sometimes the folds being even inverted; while north of it, except in its immediate vicinity, the rocks usually lie in gentle synclinal and anticlinal curves. Running across this great east-and-west fault are others that have a northward-and-southward direction. These in places shift the main east-and-west valley and the other nearly parallel valleys; therefore, it is not unreasonable to suppose that these cross-faults were formed after the valleys. These, then, are movements which have more or less affected the present surface of the ground. Furthermore, when we know that the surface of the sea has been at least 350 feet higher, relatively, than at present since the great glacial period, and afterwards has had at least one or two changes of level, it seems highly probable that there should have been some relative movement among the different beds of rocks, more especially as their continuity is broken up by various systems of master-joints.

Now, to proceed to the third and fourth propositions—if they are found to be incorrect, it follows that the fifth must be incorrect also. If the land-level is stationary, or if it is gradually rising, the natural tendency of the sea's action would be to form a plain of denudation; but if the land-level is sinking, the result will be different, for although the sea may tend to form a plain, it will not be one solely of denudation, as the marine action fills up all the low places, but does not excavate them. It would appear that the relative level of Ireland has been lowered since the historical period, as we know the time when the sea flowed into Lough Swilly, and other low places that are now bays; while we have traditions that the sea has encroached in other places, as for instance in Galway Bay, the ancient name of which is Lough Lurgan. Archæological discovery would lead us to believe that such traditions are correct, as in late years on the Aran Islands have been found human habitations and structures, which can be traced under the sands below low-water mark; while in various other places on the coast of Ireland, bogs containing trees and plants similar to those of the inland counties, are found at low-water mark of spring-tides, they being in places buried from twelve to twenty feet deep. In such places the sea does not form a plain of denudation, but if given time enough, it may form one of accumulation. Although the sea action may have a tendency to plane down all land to its own level, it cannot act so in all places, even if the land is stationary or rising. If all the rocks and different beds are homogeneous, it might so act; but on a coast like that of South-west Ireland it must operate more quickly along soft strata, while the harder rocks resist its efforts. The irregular outlines of the coast seem to have been formed in this way; long narrow bays being excavated in the places that were occupied by the softer rocks, while the hard rocks form headlands, islands, and off-shore rocks (*carrigs*, *carrigeens*, and *skelligs*). Subærialists may state that such bays are submerged longitudinal meteoric valleys; to us, however, it seems that most of the valleys of S.W. Ireland originated as bays, which are now more or less completely emergent, and that the marine action which we see to be at the present day shaping the lower ends of those valleys, is but the present continuation of that by which the upper parts of the valleys were formed.

The transverse valleys also may have been cut by marine action. At the present moment, in the south-west of Cork, the sea is excavating out the fault-rock, and forming narrow guts and ravines identical in shape with those that cross the inland

M

ridges, and form the transverse river valleys; as, for
instance, the sea is at present working across the
neck of land that separates Dunmanus and Schull Har-
bours, and across that between the north and south har-
bours on Cape Clear Island. If river action is appealed
to as having formed these parallel transverse ravines,
where did the water come from that cut the passes,
mentioned in a former chapter, through the headland
and mountain ridge south of Ballydonegan Bay? To
get a water-supply, it would be necessary that the
bay should have been once high land; and if so,
what has become of this? Rain and rivers could not
have denuded it away, and as " the internal forces
of disturbance had ceased," it could not have sunk.
As from ocular demonstration it would appear that
the sea of the present day is excavating transverse
valleys along breaks in the rocks, and as most
of the transverse valleys in South-west Cork have
been proved to lie along breaks, it seems unreasonable to say that marine action, or "the
internal forces of disturbance," could have had
nothing to do with the formation of the tranverse
valleys across the inland ridges.

The sea may act very unequally, as we have
said, along the bedding, cutting long bays into soft
beds; but when it works perpendicularly, or nearly
so, to the strike of the beds, it may denude more or
less regularly, wearing away bed after bed.   On

account, however, of the jointed and faulted condition of the rocks, it often can advance more rapidly in certain places than elsewhere; for if it meets with a dyke of loose fault-rock, or a tract of strata broken by joints or faults, it works more quickly along these, thereby at such places excavating a passage across the strata, whether hard or soft. After crossing hard beds, however, besides advancing, it will also work sideways, eating out longitudinal hollows along the soft strata. Work of this kind can be studied in Cork Harbour, the sea having cut narrow north-and-south guts through two bands of grits of Old Red Sandstone age, while in the intervening Carboniferous limestones and shales it is excavating E. and W. bays. It might be argued that none of this work should be claimed as due to marine denudation, because that when the land was relatively higher, a river may have flowed through Cork Harbour, while tributary streams, with the auxiliary meteoric abrasion, was excavating out valleys in the limestone and shale, which now are submerged, and form the lateral bays. Such may have been the case in other places, but here we know that the land is relatively sinking, and has been doing so for centuries, while at the present day the sea is yearly denuding the coast-line more and more. Consequently it is only leaving fact to go to supposition to state that a force which is not now acting probably did work which we know by ocular demonstration another force is actually performing. At the present day the sea is denuding away the soft rock, while, except in broken places, it does little work on the hard Old Red sandstones and grits. And if it is capable of executing this work now, why was it not able to do similar work in bygone ages? And the present work, if subsequently slightly modified by meteoric abrasion, would be similar to the present features of the adjoining valleys.

Cork Harbour points out the process by which the sea is prone to act, besides showing the share that wind has in marine denudation. There the sea has worked in along the principal breaks, which in all that part of the country bear about N. 20 W. The sea having breached the softer strata, and having thereby obtained access to the site of Cork Harbour, its further operations were influenced by the prevailing winds; and as these blow principally from the west and south-west, the greatest denudation was accomplished in the eastern arms of the estuary; so that now, although the sea is still carrying on its original work along the N. 20 W. line of broken rocks, yet the bay forming the harbour has been excavated principally to the eastward of that line. Similar cuttings across hard strata with associated longitudinal bays along the intervening soft strata occur in various other places in that neighbourhood; however, it is unnecessary to multiply examples, as the work in all is similar. If the land were to sink below its present level, the sea would gradually enter all the low-seated transverse valleys, as well as the longitudinal valleys, accomplishing at least some denudation as it advanced; but if the land were to rise subsequently, rain and rivers would take possession of exactly the same valleys and ravines, and commence to denude; therefore it seems to be as much a question of time and opportunity as of efficiency, which has done most of the work; since whichever held the ground longest ought to have had most effect, other things being equal. Another thing militates against meteoric abrasion, unaided by faults, having cut the transverse valleys; and that is, that the sides of such valleys are more or less abrupt, whereas this denudant usually tries to form slopes. It must, however, be allowed, as shown previously, that when this force works conjointly with breaks, abrupt steeps may sometimes result.

It has been shown that the sea is now excavating longitudinal valleys nearly similar to those that now exist in South-west Cork, allowing for subsequent meteoric abrasion and ice action; also, that it is able to cut transverse valleys, while the positions of all the transverse valleys, and probably all the longitudinal ones, were determined by systems of breaks or faults. Therefore it seems evident that the present valleys are not solely due to rain and rivers, but rather to

that action combined with glacial and marine denudation, and that all were generally led by the breaks
and faults in the rocks.

S.W. Cork, as also most, if not all of Ireland,
during the "Esker-sea" period, was relatively about
350 feet lower than at present, which is proved by
the raised beaches, terraces, and the like, found in
various places.[1]   Such a change of level would lay
most of the valleys of this country awash, and place
them in such a position from currents and tidal-
waves, that the sea would have the opportunity of
doing more work than at present.   And that the
Esker-sea did a considerable portion of the denudation seems probable, as the rivers that now occupy
the valleys since the retirement of the sea have not
been able to clear out the great mass of sea-formed
detritus, the result and evidence of marine denudation, consisting partly of Esker gravel, partly, perhaps, of estuarine deposits ; even the old terraces and
bars of the Esker-sea remain, in places, nearly intact.

Before, and during part at least of the " Esker-sea "
period, the mountains in South-west Ireland were snow-

[1] Since this was in press, our colleague, Mr J. Croll, in one of his
able papers, has shown that as the northern ice-cap advanced or retreated, the level of the sea oscillated, while the land probably remained
stationary.   This will account for the very uniform altitude of the
ancient sea-beaches; thus removing a great stumbling-block, for if the
land had been elevated, the movement scarcely could have been equal
everywhere.   Ancient beaches now found at exceptional heights, probably point to real movements in the land's surface due to local disturbances.

clad, while the valleys had their systems of glaciers,[1] a large glacier occupying the valley of the Flesk, while the moraine-drift left by others occurs in various places. The drift due to the Flesk valley glacier forms a flat-topped bank, as viewed from the Lake Hotel, Killarney ; and "this thick mass of drift conceals all the rocks along the base of Mangerton, Crohane, the Paps, and the Caherbarnagh Mountains, a distance of fully thirty miles." Here, as well as in the other valleys where glacial-drift has accumulated, rain and rivers have done very little work in re-excavating the valleys ; while in all the valleys opening westward marine action is apparent, all the glacial-drift below the 350 feet contour-line being more or less denuded and washed into gravel and shingle, the deposit of round shingle in the country north of Mangerton and Crohane being quite remarkable. In the valleys opening southward and eastward the result of the marine action is not so striking ; still we may suppose that such action took place there, as sands, gravels, and shingle are found at the lower levels, and moraine-drift at the higher.[2]

[1] "Memoirs of the Geological Survey," Ex. sheet 173, p. 7.

[2] In a valley opening south in the neighbourhood of Hungry Hill, on the north of Bantry Bay, there is unmistakable evidence of a glacier having come down within 140 feet of the present sea-level since the "Esker-sea" period. This, however, seems to be an exceptional case, and may have been similar to one of the glaciers only just discovered by Mr John Muir in some of the Californian mountain gorges—this

When rivers occur in flats bounded by cliffs, the rivers are supposed to have formed the flats. In some cases this may be right, but it is by no means necessarily so, as the sea will sometimes do exactly similar work in an estuary or fiord. The fiord called Killary Harbour divides the County Mayo from the County Galway. It is over ten miles long, and usually less than half a mile wide. The five miles toward the west bear N. 60 W.; the four next miles run nearly west and east, while the eastern part has a general bearing of about N. 60 E. It is known from the geology of the adjoining country, that the gut occupied by the fiord coincides with different lines of fault, and that each bay and turn in it are also due to faults, some of these being of immediately Post-silurian age, others of Post-carboniferous age but pre-glacial, and some probably post-glacial. It is also known that previous to its being occupied by the sea, ice flowed down it. This is proved by the presence of the moraine-drift that was left on dressed, grooved, polished, and etched rock-surfaces, which drift is now being excavated into and carried away by the sea.[1] Since the ice occupied this

glacier having occupied a similar gorge. On the north of these hills the gravels of the Esker-sea are well developed.

[1] It has been said that because a glacier flowed down this valley prior to the present occupation of it by the sea, the latter could not have made it. This, however, does not follow, as the sea may have formed it before the ice invaded the country, when the land was rising out of the water during a pre-glacial elevation of the country.

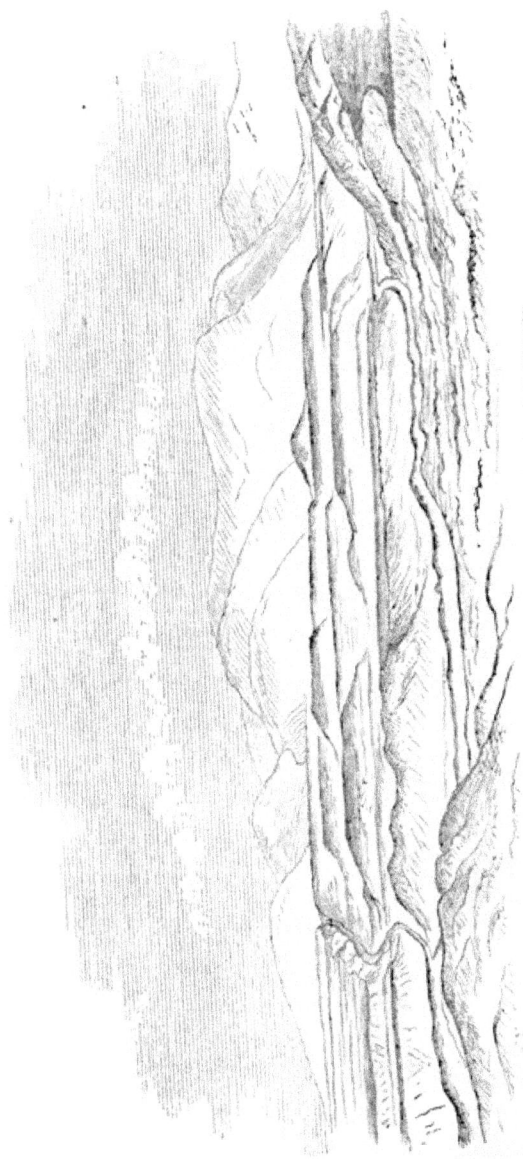

GRAVEL TERRACES SOUTH OF DERRINKEE, ERRIFF RIVER VALLEY.

valley, the land has been relatively 350 feet lower than at present. This is proved by gravel terraces, raised beaches, and shelves cut in the hill-sides, sometimes rock, sometimes glacial-drift, not only in this valley, but in the different adjoining ones. In this valley the altitude of the gravel terraces is about 350 feet, south of Derrinkee Bridge, about nine miles N.E. of the eastern termination of the present sea bay. The accompanying sketch is by our colleague, Mr J. Nolan. In Glenanane, and north of Tawnyard Lough, which lie respectively south and north of the valley, a little above the end of Killary Harbour, there are shelves or terraces at a height of about 350 feet; in an arm from the tributary valley of Fin Lough there is a flat, in Maum valley terraces, while elsewhere the traces of shelves can be detected at similar heights. There are, however, other terraces much better developed and marked than those enumerated, which can be traced from Derrinkee to Glenanane, having a regular fall from 350 feet to 175 feet in a distance of about eight miles. On account of this slope, it might be said that they must be river gravel. This, however, appears to us to be impossible, as during the "Esker-sea" period, when these gravels must have been formed, the whole of the central plain of Ireland, and the valleys in the neighbouring hills, were under the sea which reduced the present mountain-tops to islands.

Consequently there was no land for a river of these dimensions to come from ; we must therefore look for some other explanation for the sloping terraces.

We have just mentioned that at no very remote period the West of Ireland was at a relative lower level than at present, as traces of the old beaches are found in different places near the coast of Galway, Clare, Limerick, Kerry, and Cork. Since then the land must have relatively risen higher than it is at present. Afterward it again sank, as proved by the bogs and human structures submerged beneath its waters, while now it appears to be for a time stationary. If we examine Killary Harbour, we find at its east end a gravel terrace, and others occur at the mouths of the principal rivers flowing into it, while most of the rest of the old sea-margin is marked by shelves cut in the drift or rocks. If we next examine the chart,[1] it appears that, bordering the sides of the gut occupied by the bay, are sand and gravel terraces that slope from the tidal beach in the neighbourhood of Lenaun to the mouth of the harbour; while more or less near the centre there is a hollow which in general slopes seaward, the only considerable deeps being at the narrows in the vicinity of Inishbarna; while farther in are some holes connected with the junction of the lateral and the transverse faults. If, therefore, the land was gradually raised, these marginal sloping

[1] Admiralty Chart, No. 2706, Ireland, West Coast.

terraces, as shown in section (*a, a,* fig. 24, Pl. IV.), would be found at the sides of an alluvial flat (*b, b*) in which flowed a river (*c*); furthermore, the river probably would be in a supplementary flat bounded more or less by sloping gravel cliffs.[1] Advocates of sub-ærial denudation have pointed to similar flats and terraces to prove what this agent can do; as they seem to believe that only rivers can form sloping terraces. It has, however, been shown that the sea is capable of forming them; and not only them, but the supplementary flats bounded by gravel cliffs, which mark, not a river flat, but the margin of a gradually-rising estuary bottom. Even the gravels owe in a great measure their origin to sea action; as the detritus carried down into the estuary by the rivers is a bagatelle when compared with the quantity of sand and gravel that was formed by the sea from the washing and grinding-up of the moraine-drift. And similar features would be displayed by the elevation of most estuaries; namely, a terrace at the head of the valley, with minor terraces at a similar level in those places where streams or rivers flowed into the old bays, with a shelf or cliffs marking the old sea-margin; also, sloping terraces from the terminal terrace to the mouth

[1] If the land were to rise 200 feet, the terraces at Inishbarna would be about 130 feet above the water, while those at the end of the bay would be nearly 200, giving a fall of the slopes of about 70 feet in nine miles, about equal to the fall of the terraces in the upper part of the valley.

of the bay, with a river flowing in a valley, often more or less flat, between the sloping terraces. While the sloping terraces may be quite conspicuous, the marginal shelves may be more or less obscure, as they must be subsequently modified, or even obliterated in places, by meteoric abrasion.   When first these sloping terraces were observed in the valleys of Iarconnaught opening to the west, we were inclined to consider that the land had sunk more on the west than on the east; but afterwards, when such terraces were found in valleys opening eastward, we were forced to look for another explanation; and the theory we now put forward is founded on the facts recorded in the charts of different estuaries and bays, not only in Ireland, but also in Scotland.

The work done in Killary Harbour and Erriff Valley since the ice left them may be thus summarised.   As the land sank, at the beginning of the "Esker-sea" period, the sea was clearing out part of the boulder-clay from the valley, and in places doing a little rock denudation.   Part of the gravels and other detritus was carried out to sea, but a considerable portion was left behind to fill up hollows, and form a gently undulating bottom to the bay, and sloping flats in the bays and broads.   Subsequently, as the land rose, the sea partly cleared out the gravels it had previously redistributed, but left terraces as monuments of its former activity.   In the present bay the sea has been so long at work, that nearly all the old gravels have disappeared, and it is now clearing out what still remains of the boulder-clay, besides doing some rock denudation. Meteoric abrasion has not as yet been mentioned ; it must, however, have materially aided the sea, both during its advance and retreat, as it is still doing at the present day.

Here we may digress and point out the nature of estuarine deposits, and of the fossils and pebbles found in them. In some estuaries the gravels and sands will be sharp and clean, like those on an open seaboard ; but in most estuaries they are more or less dirty, clayey, or peaty, from their vicinity to the land, and the inability of the sea, confined as it is, to wash them properly ; or storms may bank up the water during floods, and cause all the dirt brought down by the rivers to be deposited inside. In Wexford Harbour, on the south-east of Ireland, during gales from the south-east, the water has sometimes been four feet higher than ordinary spring-tides. Some estuary deposits have been mistaken for glacial-drift, while in reality they are only the washing and rearrangement of such kinds of drifts. In many places on the coast of Ireland the sea is bounded by high glacial-drift cliffs which are yearly more or less denuded. In summer weather shelly sands and muds may be deposited close up to these

cliffs; while during the winter these shelly deposits
are covered over by a drift, more or less clayey,
containing numerous ice-dressed blocks and fragments.   These *glacialoid* drifts have led to innumerable mistakes; as they have been supposed, as in the
Counties Dublin and Wexford, to be a newer glacial-
drift overlying a shelly gravelly drift; while in fact
these are both members of the same drift, formed
under different circumstances, and in places graduating into one another.

If a valley were excavated solely by rain and
rivers, the gravels could only contain blocks and
fragments of the subjacent rocks, or of rocks
situated higher up the valley, pieces of which might
be carried down to any lower level. Sea-formed
gravel, on the other hand, may contain fragments
from both higher and lower levels, which are either
carried down or driven up by the waves and marine
currents; but the detritus from the upper rocks of
the valley ought to preponderate, as they would
always be coming down in rivers, or by rolling from
heights.

Estuary gravels may, in some places, contain only
fragments and pebbles of rocks from a higher level,
as a stream flowing into the estuary may supply all
the materials of which such gravels are formed.
From this it is evident that river and estuary gravel
may be nearly similar, as rivers may supply the

PLATE IV.

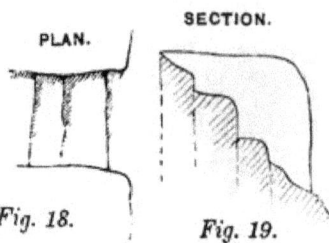

PLAN.   SECTION.

Fig. 18.   *Fig. 19.*

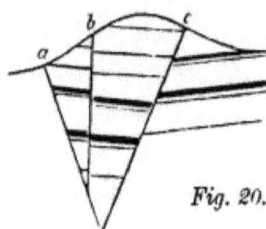

Fig. 20.

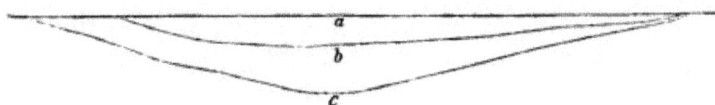

Fig. 21.

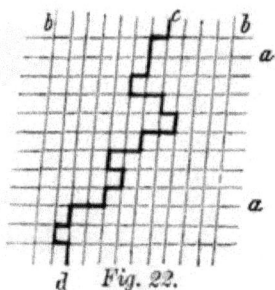

Fig. 22.

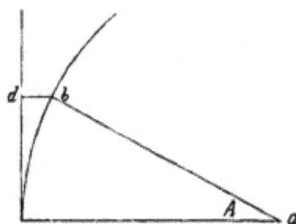

Fig. 23.

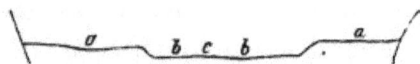

Fig. 24.

SEA CLIFF — CHALK

BASSET = GREENSAND

CHALK WITH BASE OF GREENSAND

Fig. 25.

METEORIC ESCARPMENT

SITE OF THE SEA-CLIFF

GREENSAND ESCARPMENT

Fig. 26.

materials for the latter; or, on the other hand, as estuarine deposits may contain pieces of the upper and lower rocks, and as the river gravels may be formed from them, the latter gravels may thus contain innumerable different kinds of rocks.

Estuary muds, and some sands, often contain marine shells, while gravels and peaty deposits seldom do. Terrestrial and fresh-water shells may also occur, especially in the gravels formed at the margins of the estuary. Thus, in Killary Harbour, in the accumulations at Leenaun, and at the mouths of the Bundorragha and Derrynasliggaun streams, there are more fresh-water and terrestrial than marine shells, while the bones of land animals are not uncommon; all being carried down by freshets in the rivers, and deposited in the estuary. The estuary of the Shannon, in the neighbourhood of Limerick, formerly extended much farther than at present, it having been margined by mud-flats, corcasses, swamps, and salt-marshes, which are now reclaimed, some within the last few years. These evidently were estuary deposits, formed very recently; as before they were embanked they were covered during every spring-tide. In some places marine shells are very common; but associated with them are the horns and bones, and sometimes even entire skeletons, of the elk, reindeer, red deer, &c. Some of these bones were evidently carried down into their present position by the waters of the Shannon, while others probably are the remains of animals swamped while feeding on or crossing the marshes, in the same way as horses or cattle may be swamped at the present day. Land and fresh-water shells ought necessarily to occur more frequently in the sands near the mouth of a river than elsewhere; but they may also be carried out to the terrace gravels. Professor Hennessy, F.R.S., &c., has pointed out the conditions under which sand and portions of shells will float.[1] On a warm and dry day, the dried shells and sand, when the tide rises to them, repel the water, and swim thereon, and as the tide moves gently past, they rise one by one, and float off. We have seen shells floating also, after the manner of boats; but the sea must be perfectly calm, as the moment they meet a ripple or breeze, they are over-balanced, fill, and sink. In this way land shells, although so frail, may be carried out and deposited in salt water.

The sea in an estuary is quite capable of levelling alluvium, even although coarse shingle, and thereby forming extensive flats; this, however, is not the case with a river, as the materials carried by it, especially if coarse, are usually piled in banks. A river-bed may be prolonged forward, by the stream carrying gravel, &c., into a lake, forming a level

[1] "Floatation of Sand," *Geological Magazine*, vol. viii. p. 316.

(*srah* or *haugh*), and eventually a plain bounded by
cliffs; but in this case neither the flat nor the cliffs
are due to the river action, but to wind combined
with the water of the lake. True, river-flats may be
formed by rivers gradually eating into a hill on one
side, while the debris on the other side is levelled
during floods; such flats, however, are of small
extent, and in Ireland all the extensive river-flats are
differently formed. When the rivers flow more or
less near the centre of their flats, they add yearly to
the alluvium, quantities of silt and such like substances that are suspended in the water during floods,
but never gravel and sand; except immediately below
rapids, where the coarse materials can be driven on to
and scattered over the plain. Irish rivers, therefore,
may be related to the flats through which they flow in
four different ways :—1st, They may in some places
form their own flats and the marginal cliffs; 2d, They
may fill a lake-basin with gravel, and form a flat, but
not the marginal cliffs; 3d, They may occupy old
valleys and raise their floors by annual deposits of silt;
and 4th, They may flow through old estuary-flats,
to which the marginal cliffs were formed previously.
In the first case, rivers would take ages to form a
flat of any extent, and to lower their beds subsequently as the land rose; during which operation they
would probably leave behind upper-river gravels in
the valleys. In the second case, rivers might possibly

N

form upper gravels; for as the land rose and the *embouchures* of the lake-basins were down, the rivers would partly re-excavate the gravels they had previously deposited in the lake-basins. In the third case, the flats may or may not be solely due to the rivers. When sections are opened across them, it is often found that at the first they were a bog or morass, and that vegetable growth and decay formed the beginning of the flat, while subsequently river silt was deposited over the peat. In other cases will be found a substratum of gravel or shingle, probably an estuary formation, on which the river alluvium has been deposited. Such rivers, however, might easily lower their beds if the land rose, and the barrier across the lower end was capable of being easily denuded. But in the fourth case, rivers have little or no work to do; for prior to and during the rising of the land from under the sea, the flats and terraces, with the upper and lower gravels, would be formed.

CHAPTER XIII.

FROM the description in the last chapter of Killary
Harbour, and the different kinds of estuary deposits,
it is evident there are no proofs that the flats in the
County Cork below the 350 feet contour-line were
formed by the rivers flowing in them, while it
appears highly probable they may be old estuary
bottoms, as the gravels associated with them are at
certain elevations in all the different valleys, as if
due to one universal agent, such as the sea. In some
of these flats the work of rivers is apparent; these
having cut passages here and there, and afterwards
filled up the channels, or perhaps left them as irregular trenches after they had taken new courses.

From S.W. Cork we now proceed to the valley
of the Weald, which has been put forward as one
of the great proofs of enormous subærial denudation. We will specially refer to Messrs Foster and

Topley's paper on it, as these observers have carefully examined the geology of the country.[1]

The Weald is bounded on the north, west, and south by chalk-hills or Downs, while near its central line is an axis of hills, which in places is higher than the average height of the Downs. These central hills are of rocks the oldest in the area; consequently they are hills of elevation, which, however, have been greatly modified by denudation. If on the rocks forming these hills the absent strata were replaced, the hills would be between 2500 and 3000 feet in altitude, or in round numbers, about 2000 feet higher than at present. The North Downs are, on an average, 700 feet high, and the South Downs 800 feet; consequently the central hills would be 1300 and 1200 feet higher than the present North and South Downs respectively. These are about 18 and 12 miles from the central axis, and there are dips northward and southward (ignoring all minor flexures and rolls) of about 78 feet and 100 feet to a mile, equal respectively to about 1 in 68, and 1 in 53, or about an angle of one degree both ways.

If subaerial agencies are as powerful as some suppose, it is quite unnecessary to call in any aid from marine denudation, and it may be supposed

[1] "On the Superficial Deposits of the Valley of the Medway, with Remarks on the Denudation of the Weald," *Quarterly Journal of the Geological Society of London*, November 1865, p. 443.

that this agency has removed at least 2000 feet in thickness of strata over all the area now occupied by the Hastings beds, a less thickness over the Weald clay area, and still less, in different degrees, over the country now occupied by the greensand and gault. Not very long since, the Hastings beds and Weald clay areas was a forest, the growth of which must have prevented all surface denudation; and during this Forest age the rainfall must have been a great deal more than it is now.[1] Although the Hastings beds and Weald clay areas were protected by their envelopes of trees, the surrounding chalk downs must have suffered considerably, which seems proved by " the beds of unstratified flint-gravel that are met with in many places on the lower greensand, gault, and lower slopes of the chalk."[2] Now, however, although the rainfall has decreased as the forests are gone, atmospheric denudation is acting more or less over the whole area, being accelerated in many places by the tillage of the land.

The extreme advocates of subaerial denudation may fairly be asked:—1st, Why did the forest grow, and stop the denudation of the Hastings beds and the Weald clay? 2d, Why are there not accumulations of chalk flints on the Hastings beds and the Weald clay areas, as well as on the gault and greensand?

---

[1] See page 113.   [2] Foster and Topley, p. 446.

And 3d, How the subærial agencies at the first formed an escarpment in the chalk?

To consider the last first: subærial agencies, unaided, seem incapable of cutting a ravine in homogeneous chalk or limestone. We do not know whether a bare sheet of chalk exists, but many such of limestone do, and in all cases meteoric abrasion denudes the whole surface; so that no ravine, and consequently no escarpment can form, except the rock is faulted, jointed, or traversed by some kind of shrinkage fissures, which are acted on by the weather. But as in such rocks master-joints and faults are nearly always connected with an underground drainage,[1] they can seldom be the forerunners of valleys that are denuded solely by meteoric abrasion. Once a ravine or valley is begun, whether it be due to the shrinkage of the rocks, an anticlinal curve, marine denudation, or the scooping out of ice, meteoric abrasion would be capable of materially assisting in developing it; more especially if the beds dip away from the valley, and some are softer than the rest, as the basset of the soft beds will be denuded more quickly, and cause a hollow in which water will collect, and form a stream at the base of a cliff; the water conveying off the debris as fast as it falls. If, however, the beds dip into the valley, the water will run away from the cliffs, and all debris that falls will remain at their

[1] Whitaker, "Memoirs of the Geol. Survey," Sheet 7, England, p. 96.

bases, till eventually slopes are formed; on this account valleys usually move away from their faults, the rocks on the down-throw side, which usually rise to the fault, being more easily denuded than those on the other side. Wind, however, as previously pointed out (page 80), may act very similarly to a stream of water in keeping a cliff perpendicular. Joints or other shrinkage fissures will also affect the weathering of a cliff, according as they are nearly parallel or perpendicular to the line of cliff, as previously illustrated when speaking of the marl cliffs in south-east Ireland (page 51).

In the Weald there would appear to have been at one time more or less continuous cliffs to the north, east, and south of the clay country, probably at the commencement of the Forest period; these, however, seem to have been denuded back since then, and modified by meteoric abrasion, which first changed the cliffs into slopes, and afterward gradually denuded the latter backwards; so that now the escarpments may be far removed from the original sites of the cliffs, and during all this time the transverse valleys across the Downs must have been somewhat similarly acted on. If the sea had formed cliffs to the Weald valley, it would necessarily have worn back both the chalk and the harder underlying strata; such, however, would not be the case with meteoric abrasion, as this agent would have greater effect

on the chalk, forming a continuous escarpment of
that rock, but leaving behind it prominent masses
of greensand, to mark, but in places only, the site of
the ancient sea-cliff. From this it is apparent, that
although the escarpment is now being gradually
worked back by meteoric abrasion, yet it may have
been due originally to sea-work. The accompanying
sketch maps (figs. 25, 26, Pl. IV.) will explain how
this may happen. Fig. 25 represents a marine cliff in
one place cut entirely in chalk, and in another part
formed of chalk and greensand; while fig. 26 represents the present chalk and greensand escarpments,
part of the latter being the base of the old sea-cliff,
while the sea-cliff that was entirely chalk has been
totally denuded away. If the Weald valley was solely
due to subærial denudation, there ought to be deposits of chalk flints over the whole area, and not
only on the newer beds. If, however, the sea formed
the original valley, it would have been capable of
carrying the flints out of it, and consequently flints
would only be found on the beds exposed by meteoric
abrasion, since the sea ceased to act. Furthermore,
when the sea retreated, the forest grew, and occupied
the Weald clay and Hastings beds areas, protecting
them; but meteoric abrasion was still working on the
chalk, denuding it backwards, and exposing the green-
sands and gault.

It must be allowed that, at one time, there was

a horizontal mass of chalk over all the Weald, and
that consequently the North and South Downs sank,
or the Hastings beds were shoved up. It has already
been shown, that when a sheet of rock is bent into
anticlinal and synclinal curves, it may form unbroken
arches, or the strata may gape in places, and form
open fissures along lines of weakness, the gaps being
equal to the radii minus the cosines of the angles
of the slopes. This, in the case of the Weald, if we
ignore the minor flexures, and consider it one anticlinal,
would be 18 miles multiplied by 1 minus the cosine of
$1° + 12$ miles multiplied by 1 minus the cosine of $1°$,
which would give a valley only 25 feet wide from
escarpment to escarpment, if the beds opened together, which, however, is improbable, the different
groups of rocks being made up of such dissimilar
materials. In such a case, each bed would have re-
tained its relative position to its fellows ; but if the
beds were capable of bending without breaking, and
would not stretch, the original position of each would
have to change, as each bed must move on the one
below it up towards the axis of each anticlinal curve.
If, however, the upper beds were ruptured, while the
lower beds were not, it is not unreasonable to suppose,
especially if there was an upward as well as a horizontal thrust, that the upper beds might remain
fixed at their unbroken ends, while the lower beds
would be pushed along below them, and up into the

break, such a movement naturally taking place along
a soft, weak bed. Thus, if the chalk in the Weald was
broken, but fixed at one side, the greensand would
be pushed out from under it along the gault; and the
Hastings beds might be pushed out from under all,
along one of the Weald clays. Such a movement in
the Weald, although the average angle of dip be only
one degree,[1] would form terraces between the outcrops
of the chalk and greensand, and the greensand and
the Hastings beds in the north and south parts of
the valley respectively, 15 feet and 10 feet wide.

As soon as this movement was complete, or even
while it was yet in progress, denudation would be
at work—marine denudation if the movement was
submarine, meteoric abrasion if it was subærial. Thus
either denudant, if it had time enough, might possibly
have accomplished the work that has since been done;
but it appears much more natural to suppose that
neither of these agents did all the work, but that
their efforts were combined; each accomplishing its
allotted portion. Le Conte has shown that it is probable that all elevations and crumbling up of the
earth's crust commence beneath the sea, and all
rocks thus raised must necessarily be exposed to the
full force of the sea prior to being raised above its
influences. In the case of the Weald, the sea would

---

[1] The average angle is higher; this is the lowest mean angle of dip
that can well be assigned.

at the first remove all broken portions from the central axis; afterwards, as the land gradually rose, it would denude back the chalk, and also form transverse valleys along the different secondary breaks. The land may even have had various upheavals and lowerings, at one time being above, and another time below the sea, and during all such movements it would be exposed to denudation by the different forces, and this appears probable from this, that most of the rivers now flow through the transverse valleys; but during the different risings and lowerings, such places would be liable to be deepened by the sea, more than the longitudinal valleys; which is the reason why, when the land finally rose, they would form the channels most likely to be occupied by the rivers. Such an estuary would also account for the marked escarpment bounding the valley on the north, west, and south, while none exists bounding the high land in the centre, for the sea would form cliffs in the chalk as the beds dipped away from it, but as the beds in the central axis dipped towards it, the sea could only form slopes, it being unable to work back the base of the escarpment as fast as meteoric abrasion was wearing away the upper portion.

It has been intimated that breaks do not exist in connection with the transverse valleys of the Weald; but it would appear to us that a fault must occur in the valley occupied by the River Ouse; while between

Lewis and Brighton, we traced numerous breaks, not only forming features on the escarpment, but also across the Downs, to which are due the irregularities in the boundary of the chalk. Furthermore, Messrs Foster and Topley's suggestion would corroborate the idea that these ravines and hollows are connected with breaks, although these breaks are not marked on the map; for they state, " These valleys are probably due to the dissolving away of the chalk along lines of underground drainage." Mr Whitaker also has suggested the same, in relation to the ravines in the chalk of other places. There are also the ponds on the Downs, which must be connected with subterranean streams (similar to the turloughs in the limestone districts of Ireland), and become filled during wet weather (the underground passages being too small to carry off all the water), but are dry at other times.

The authors of this paper on the Weald seem to believe that in landlocked bays sea action is necessarily feeble. This, however, does not appear to be borne out by observed facts, as after the high tides of 1870 we found on the west coast of Ireland, that on the open seaboard little work was done even on such frail materials as those forming the sand-dunes, while in the bays and other landlocked places, cliffs, piers, embankment roads, and everything the tides could reach, were more or less injured or carried away. On an open seaboard the sea can do little work, unless it is assisted by wind, but in narrow places the rush in and out must accomplish more or less work every tide.

It seems remarkable that the gravels of the Weald, if fluviatile, should all be below the 350 contour-line, like the gravels of the " Esker-sea " period in Ireland, and the sea-gravels in North America.[1]

The gravels of the Weald are said to be fluviatile ; no positive proof for this suggestion has, however, been given ; and we have shown that many so-called river-flats and river gravels are probably the ancient bottoms of estuaries and the gravels of estuaries ; as in an estuary sloping terraces of gravel become formed, while terrestrial and fresh-water fossils may not be uncommon ; for estuary deposits are more likely than river accumulations to have the bones of land animals in them, as may be studied in any river during floods, because all animals drowned are swept down by the current, and their bodies deposited in the still water of a lake or an estuary. If there are rivers flowing into an estuary while the waters of the estuary are denuding rocks, the gravels found in the estuary will be composed of fragments of higher beds brought down by the rivers ; while the local fragments will be brought elsewhere according to the set of the tidal and other currents ; and such gravels could not be

[1] Principal Dawson.

regarded as purely fluviatile. The coarse and fine deposits in the drift of the Weald valley would appear to be estuary accumulations, as similar series are found in all estuaries—in one place nothing but clays and the like, and in others gravel or shingle; while rivers must form alternations of fine or coarse accumulations, according to the height of the waters.

Proofs are not wanting of the valley of the Weald having oscillated up and down during the most recent ages. According to Mr Drew,[1] a bare pole was put down 70 feet in Romney Marsh without reaching the bottom of the recent accumulations, the lowest 50 feet of which were sand containing "recent marine shells, especially cockles." Over this sand there are in places five or six feet of peat, containing the roots, stems, and fruit of trees, such as the oak, alder, hazel, &c., and over the latter a clayey, peaty, or sandy alluvium. Clay also sometimes occurs overlying the peat, while "the whole of the alluvium is below the level of high water at spring-tides."

According to Dr Mantell,[2] "in Pevensey Level the trunks of large trees have often been observed imbedded in a mass of decayed vegetables. The substratum is an inferior peat, with an intermixture of sand reposing upon a thick bed of blue alluvial clay, containing marine shells;" while at Irfield, "nearly

---

[1] "Memoirs of the Geological Survey, England," Sheet 4.
[2] Mantell's "South-east of England."

twenty feet" of peat, containing tree remains, was
sunk through, and in various places off the coast
submarine forests are known to exist. According to
this authority, the general section of the low levels is
as follows :—

4. Vegetable mould.
3. Peat with trees,   .   .   .   3 to 5 ft.
2. $\left\{\begin{array}{l}\text{Dark-blue silt (clayey), with}\\ \quad \text{fresh-water shells above and}\\ \quad \text{marine shells below, .} \quad .\end{array}\right\}$ 5 to 25 „
1. Pipe clay, ...   1 to 4 „

34 ft.

From the above facts it is evident that the land
once stood much higher than at present, during which
time a bay over 70 feet deep was excavated in the
place now occupied by Romney Marsh; that afterwards the land began to rise higher, and the bay was
filled with estuary deposits. At the first the sea had
free egress into the estuary; but afterwards it became
a fresh-water lagoon, and eventually dry land, on
which a forest grew and flourished. But subsequently
the land again began to sink and be subject to floods,
while mosses and the other peat-forming plants
established themselves. The trees then began to
decay away, till eventually all were destroyed and
covered with peat. After that the land sank so low
that even the peat could not grow, the place being con-

tinually flooded with water, and during this time the clays, alluvium, &c., were deposited over the peat. Because a bog containing trees, &c., may be found inside and outside a sea-beach, it has been argued that the trees may have grown at their present level, while since then the sea has moved its beach in landwards, exposing the bog and trees to seaward. The fallacy of this reasoning is easily exposed if we only examine into the conditions under which the trees found would grow; as oak and most other trees cannot be grown except on drained land, which could never exist naturally in places below high-water mark.

CHAPTER XIV.

THE LOCH LOMOND AND OTHER SCOTCH VALLEYS.

Scotland, similarly to Ireland, is intersected by systems of breaks, but they are in general more intricate and less regular; this resulting apparently from their being so numerous and of so many different ages, the newer deflecting and displacing the older. However, in some localities in Ireland, such as N.W. Ulster, W. Connaught, and S.E. Leinster, faults and breaks are just as numerous and complicated. In the Highlands of Scotland, as far as we visited them, we did not meet with a valley, ravine, or lake-basin unconnected with a break. It would also appear that in most cases there is probably a connection between ice action and the lake-basins; but this denudant seems to have been always led by the faults, breaks, and other shrinkage fissures in the rocks. If it is admitted that lake rock-basins are partly due to ice action, and partly to the contraction and the displacement of the rocks, all objections now raised against the ice theory would be done away with; as the hollows at the outset would be formed by the crossing or junction of breaks, while subsequently ice action would lift up and carry away

o

all the loose blocks and fragments, thereby clearing out the basins.

Loch Lomond occupies a remarkable valley, which has a general bearing of north and south, corresponding with the lie of other important breaks in that part of Scotland. Extending from this valley eastward and westward are transverse features—some forming valleys, while others constitute greater or less depressions in the hills. That all these valleys and depressions are connected with dislocations in the underlying strata is manifest by the strike of the rocks being deflected, or the beds of rock rising to the up-throw of the faults, or by the shifting of conspicuous beds of rock.

Loch Lomond must, at no very remote period, have been a fiord connected with the valley of the Clyde. This is proved by the raised beaches of the latter valley being on a similar elevation with the terraces in the Loch Lomond valley, and the gravel-bar across the latter, a little to the north of the lake. The lake-basin, therefore, at that time, was a submerged valley, which prior to its submergence, if the theory of the subærialists be correct, was excavated by rain and rivers, or ice.

The lake is divided into two portions by the chain of islands stretching westward from Balmaha. From the chart [1] we learn that south of these islands it is

[1] Admiralty Chart, Scotland, Loch Lomond.

shallow, rarely exceeding 12 fathoms in depth; however, in a few places it is 13, and in one spot, east
of Inchmurrin, a hole 14 fathoms deep is recorded.
North of the islands the lake gradually deepens to
Ross Point; it then shallows for a short distance,
being only four or six fathoms deep on Hunter's
Bank, which lies a little north-east of the inver of
Douglas Water; but on the north of this bank it
immediately deepens to over 25 fathoms, and northward it gradually gets deeper and deeper, till it
attains the maximum depth of 105 fathoms due
west of the hamlet of Culness.   North of this place a
deep portion (over 100 fathoms) extends for nearly a
mile; after which the basin gradually shallows to
Whitepoint, between which and the end of the lake
there is another deep portion, about 34 fathoms.

From an examination of the country in the neighbourhood of the south part of the lake, we learn that
there are features connected with breaks running
about N. 30 E., while the deeps in the adjoining
portion of the lake-basin have similar bearings.   Immediately north of the islands the direction of the
deeps is not so regular, they in part running N. 30 E.,
and in part nearly N. and S.; but in the rest of the
lake the deeps extend nearly N. and S.   It may be
observed, that although the general bearing of the
lake is N. and S., yet the deep portions in the north
part have not an exactly similar lie; as in places

they incline more or less to either side, from being
deflected at the junctions of the transverse valleys
with the main one.   All the deepest spots coincide
with one or other of these junctions.

The shallows in the lake may possibly be due, in
part, to accumulations made when the lake was a fiord;
similar shallows, however, are not found in the fiord
of the Clyde, and it is probable that since the lake
became fresh water they have been considerably augmented by debris carried on to them by the rivers
and streams, especially the bank in the vicinity of
the Douglas Water, which has all the character of a
delta, shelving suddenly at its margins, especially
northward.

As already stated, the features of the adjoining
country and the shapes of the bottom of the lake-
basin have a connection between them.   The irregular deep, east of Ben Dhubh, occurs where the
nearly N. and S. breaks and the N. 30 E. breaks
meet or cross one another.   The sudden deepening
north of Hunter's Bank seems due in part to a fault
with a down-throw to the north, coming into the main
valley from the east.   The shallow to the south of
the deepest spot is evidently on the N.W. side of the
fault-line which is associated with the Tarbet valley.
This appears to be a very recent fault, and it has considerably shifted the main fault of the valley.   The
Inversnaid and Inveruglass faults also slightly shift

the main fault; while the deepest spot in the basin
(105 fathoms) seems to be at the junction of the main
and Culness valley faults. We have now stated the
principal facts in connection with this lake-basin, and
proceed to offer suggestions as to its formation.

If this valley were solely due to rain and rivers, it
would necessarily deepen gradually from its upper to
its lower or south end; this, however, is not the case.
If due to the erosion of ice, the deeps and shallows
ought to graduate one into the other, and have a
trend conformable with the lie of the valley, and
consequently with the course of the glacier; this,
however, is contrary to facts. Again, marine action
could not possibly have excavated this valley; on the
contrary, it would tend to fill it up; but the extreme
shallowness of the lower or south portion seems not
to be due to such a cause, as the soundings record a
rock bottom in many places. It should, however,
be pointed out, that in some of the neighbouring
fiords, such as Loch Fyne, there are extraordinary
deeps in the upper portions, which apparently are not
being filled up by sea action.

It has been shown that the valleys and depressions
at the surface are connected with breaks in the strata,
while the bearings of all the deeps in the lake-basin
are similar to those of the features in the adjoining
country, or they have a relation to the several junctions of the transverse valleys with the valley of Loch

Lomond; consequently, we have a right to assume
that the form of the lake-basin is more or less connected with the breaks in the underlying strata.   Let
us suppose a great north-and-south break to have
been first formed, with transverse breaks and their
accompanying fissures branching from it.   Meteoric
abrasion, or if under the influence of the sea, marine
action, would widen and deepen the fissures or other
vacancies that were thus formed; or if the climate
became arctic or alpine, ice would catch up and carry
away all loose blocks, forming valleys along the fissures and lines of breaks; while in the broken ground
at the crossing or junction of one or more of these lines,
the ice would wedge up and bring into the transport-
ing power of the glacier all loosened blocks and fragments of the rocks, and consequently excavate the
rock to a great extent, thus forming the deeps; then
subsequently produced breaks or dislocations would
shift the work previously accomplished, and more or
less modify it.

By operations such as those mentioned it seems
possible to account for all the phenomena of the basin
of Loch Lomond; ice having accomplished the major
portion of the excavation along broken lines of rock
due to faults and other shrinkage fissures.   For the
formation of the deep spot due west of the village of
Culness a second theory may be propounded; for
when the land was relatively higher, and the deeper

portions of what is now the lake-bed were above the level of the sea, underground vents may possibly have existed, as we know they do in other parts of the world.

Suppose the country about Loch Lomond to be 600 feet higher than at present, and rain and rivers to be excavating the basin. This hole being at the junction of the main and Culness valley breaks, it is possible that in this place there might have been a subterranean outlet, which was capable of draining the water out of the greater portion of the basin of Loch Lomond. Such subterranean outlets at the junction of two or more breaks are not uncommon among similar rocks in the hills of Iarconnaught, Ireland, where the "fault-rock" is of an incoherent nature or open structure; or when an internal fissure has not been filled with minerals. Such a theory as this would necessitate that all the rock debris taken out of the valley—which may have been more or less in accordance with the size of the original fissures due to the shrinkage of the rocks—was carried away through the subterranean passages; now we have previously shown, when describing the subterranean drainage in the neighbourhood of Gort, County Galway,[1] that a considerable amount of detritus, and even blocks of stones, may be carried away through such passages. This theory, however, does not appear satisfactory; for although in most lake-basins, especi-

[1] "Memoirs of the Geological Survey, Ireland," Ex. sheets 115, 116.

ally at high levels, it is probable there is or was a certain amount of subterranean drainage from them, yet it seems scarcely likely that it could have accomplished much of the excavation of the basins. We are therefore inclined to believe, that after the main lines of the valley were laid out by faults or other shrinkage fissures in the underlying strata, marine action, meteoric abrasion, and ice, separately or conjointly, excavated the basin, that during this process other shifts occurred which modified its original form, while subsequent shifts or shrinkings helped to form the remarkable deep holes. It may also be suggested that among the later movements of the surface, the lower portions of the lake may have been lifted higher than the northern part; or the upper portion may have been depressed more than the south part. That some of the dislocations in connection with this valley are post-glacial, we are strongly inclined to believe. Our examination, however, was too brief and restricted to justify us in coming to a positive conclusion on this point; but the features of the country about the north part of Loch Lomond, and also in the neighbourhood of Loch Katrine, seemed to indicate very recent movements and breaks in the underlying rocks.

Before leaving this lake-basin it should be pointed out, that if at the time it was a fiord there was an arctic climate, as the fossils found in the raised beaches

of the Clyde valley would indicate ; the fiord may have
been covered with ice, or have had an "ice-foot," on
which was caught most of the debris brought down
from the hills to be floated out to sea (little or none
of it remaining to be deposited in the lake-basin to
silt up the deeps), and if the ice remained longer in
the upper portion than in the lower, the former would
have been more excavated. These suggestions would
apply to the neighbouring fiords also, and to some of
the bays in the north of Ireland.

Our colleague, Mr H. Leonard, M.R.I.A., thus
writes of some of the Scotch valleys:—"In the
Highlands of Scotland I was greatly struck with
the *form of the ground*, the resemblance to that of
portions of Connemara being complete. The rocks
are similar in composition, and the physical features could hardly fail to command attention, while
the absence of timber increases the opportunities for
observation. The faults here, as in the west of Ireland, run principally through the *cols* or *maums*, that
of Glencroe passing through the maum at ' Rest and
be thankful.' The historic valley of Glencoe lies
along a line of break in its schistose rocks, which is
very prominently marked from about the centre to
the top of the glen, where the main fault appears to
split into a number, and these are in many cases cut
across by other faults. The glen of the Blackwater
Lochs is another example of a valley along a break.

In this wild and rocky gorge the strata are well exposed, so that the data are easily observed. The bearing almost coincides with that of Loch Leven, in which the waters of the glen flows. The Caledonian Canal, which consists of a series of connected lakes, is also in the line of a great fault, and very many cross-faults were noted along its route.

" Loch Awe, from the nature of the rock, does not as plainly tell its tale ; but the facts observed in the vicinity of Portsonachan, and along its western side, rather incline to the assumption that it is connected with shrinkage breaks ; the joints having nearly the same general bearing, and indicating that two sets of master-joints aided in producing the physical features here developed. Breaks or master-joints appear to run into the two glens at the head of the loch.

" In the Grampian Mountains the resemblance to Connemara is, if possible, more striking, as the country is very much faulted, like our own ; and I cannot help thinking that any observer visiting the district, supplied with the key obtained from a lengthened examination of a similar region, may more truly unlock their history, and, perhaps, cease to be ' fully persuaded that these valleys are to be looked upon as the results, not of subterranean movements, but of subaerial denudation.' I may mention, that when travelling through Savoie, Valais, and the Bernese Oberland, I was greatly impressed with the numerous

examples presented of valleys excavated in the line
of faults.   The Arve, Rhone, Visp, and their many
secondary valleys, show countless instances where
subterranean movements have aided denudation in
carving out the features now existing, and in the
narrow rocky gorge where the Aletsch glacier (the
longest in Europe) ends, the work of denudation,
combined with shrinkage lines, is most apparent.
That both agents should be allotted their respective
tasks appears to me necessary in unravelling the
geology of most districts."

An examination of the Chart of Loch Fyne[1] is not
without interest.   The south portion of this fiord, as
far north as Loch Gilp, bears nearly N. and S., but
the rest of it has a general bearing of N.E. and S.W.;
the fiord widening out considerably where these two
portions join into one another.   When, however, it
is examined in detail, other peculiarities will be
observed.   At the entrance of the bay, and south of
a line (N. 70 E.) drawn through East Loch Tarbert,
the deeps run about N. 15 W.; but at this line they
are shifted considerably towards the west, while north
of it they bear nearly N. and S. to High Rock, which
lies due N.E. of Maol Dubh Point.   A little S.E. of
Maol Dubh Point there is a space over 80 fathoms
deep, with one portion over 90 fathoms, and from
this deep there extend others of less magnitude

---

[1] Admiralty Charts, Scotland, West Coast, Loch Fyne.

towards N. 20 W. into Loch Gilp, and N. 15 E. to the
Otters Spit Narrows, the 90-fathom hole being at the
junction of these two lines.   Immediately north of
the Otters Spit Narrows the line of deeps is shifted
considerably from the west towards the east; while
north of this line the deeps bear N. 20 E. till they
meet the break (N. 20 W.) coming down Gair Loch,
at the junction with which there is a deep hole, over
30 fathoms deep, the surrounding bottom being less
in depth.   Also starting from this point is another
line of deeps which bears N. 40 E., and extends to
the Minard Narrows.   At the Minard Narrows there
is a complication, there being various small islands,
and only one narrow channel, which reaches 20
fathoms in depth.   North of these islands the line
of deeps is slightly shifted towards the west, and
it bears about N. 30 E. as far north-east as
Furnace, when its bearing changes to N. 70 E.;
but in a short time, when Strachur Bay is nearly
reached, it changes to N. 25 E., and this line of
deeps extends past Inverary into Loch Shira.   From
a little east of Furnace to Inverary the fiord is very
deep, most of it being over 50 fathoms, while there
are long holes in it over 70 fathoms deep.   From
Inverary toward the N. 60 E. the deeps extend for
some miles, while the channel of the fiord at the
north-east extremity bears nearly due N.E.

From the summary just given it is evident that

the fiord of Loch Fyne consists of a series of features which are more or less systematic; for the deep portions extend along regular lines, and where these lines cross or join into one another, there are extra deep soundings, and each shift in these lines corresponds with lines of features in the adjoining country. Glenshira is connected with a break in the strata, and Glenary appears as if it followed a similar line. However, the drift greatly obscures the evidence in the latter case; but the sudden deep opposite Inverary lies at the junction of these two lines with those of the breaks that exist higher up and lower down in the main channel of the fiord; while a little farther S.W. one of the deepest spots in the upper portion of the loch is where the break occupied by the Douglas Water joins into and slightly deflects the main break. The island and shoals at Minard Narrows seem to be brought up by a fault; so also does the Otters Spit, and the various islands and shoals in the bay; but as the country in the vicinity of Loch Fyne was not all examined, nothing further can be said about them.

The lower portion of this fiord has a nearly gradual slope from the Frith of Clyde to the Otters Spit Narrows, but N.N.E. of these to Minard Narrows there are deeps, which, however, are not as considerable as those previously mentioned in the reach between Minard and Loch Shira. These different soundings

may be accounted for as follows : the lowest portion
is gradually filling up by sea-brought detritus ; the
centre portion is also slightly silting up ; while the
sea has very little power in the upper portion.
Furthermore, ice may have existed longer in the
upper than in the middle portion, in consequence of
which, the former would be more worn and hollowed out thereby, than the latter, while most of the
detritus brought down would have been caught on,
and carried away by, the ice, so that little or none
of it could now remain in the upper part of the basin
of the fiord.

APPENDIX.

FORMATION AND GROWTH OF SOIL OR SURFACE-MATTER.

WHEN land is first exposed above water, or from under ice, it is devoid of soil. Soil or vegetable mould may be earthy, sandy, or organic; it takes years to form, but it increases much more rapidly under some circumstances than others; the process being due to the combined effects of meteoric action, the growth and decay of vegetable matter, and the operations of certain animals.

All rocks, no matter how hard and tenacious, will, in time, weather and disintegrate if exposed to meteoric action; but the detritus formed may not always remain to produce soil, for if the land slopes sufficiently, all, as fast as formed, may be removed by rain or wind, the latter force being even capable of removing detritus from a flat surface, especially if in exposed situations. Many rocks, such as doleryte, dioryte, granyte, some limestones, &c., are physically hard, while they are chemically soft; others, such as slates, schists, clays, &c., are physically soft, while they are very little susceptible of chemical decom-

position ; therefore on rocks of the first class, soil
will form much more rapidly than on those of the
second. Some rocks, however, may weather easily
into a meteoric accumulation, and yet, on account of
a want of a necessary combination of substances, will
be incapable of forming a productive soil. Some
granytes and sandstones thus weather into "cold
soils;" but if the former be associated with limy
matter, and the latter with marly or clayey stuff, a
"warm" or productive soil is the result. Other
rocks are so chemically formed, that as they
weather nearly the whole of their constituents will
be carried off in solution. Thus in some parts of the
County Galway, over the carboniferous limestone all
the limy matter has been carried away through
swallow holes or "sluggys," leaving a siliceous surface accumulation, due to the decomposition of the
contained chert. In some low-lying places this
sandy drift may have been carried down from the
adjoining high land ; but when we find it occupying
high land, it must have been formed where we find
it.[1] Many earthy limestones weather into valuable
soils. This, however, depends very much on the
nature of the climate; for while in a moist, cool
country the land over limestone may be most fertile,
in a dry, hot country the land on an exactly similar
rock may be unproductive.

"Memoirs of the Geol. Survey, Ireland," Ex. sheet, 107, p. 34.

Boulder-clay drift has in general been formed out of the denudation of various rocks; and we find fertile soils more often over it than in places where the surface is due to the disintegration of only one kind of rock. There are, however, exceptions to this rule; as some rocks, especially if belonging to the Kainozoic epoch, may be so constituted that they form excellent soils. Drift, moreover, may be formed almost solely of one substance, such as clay or sand, when it will weather very slowly; or it may contain subordinate layers or beds of an impervious character, such as layers of bog iron-ore, or beds saturated with other minerals, which will retard, or even prevent, the growth of soil in places, or even over large tracts. In many of the estuaries of South-east Ireland a sandy marl has been deposited, which, if exposed to atmospheric influences, would freely weather into a rich soil; this, however, in most places is covered by a ferriferous clayey or sandy deposit, that will not weather freely or support vegetation. The good marl was evidently deposited while the water was deep, all the iron brought down in solution by the streams from the adjoining country being carried out to sea; but when the estuaries began to fill up, and the muds to be exposed twice daily, more or less of the water evaporated, depositing the iron it contained. In the north mud-lands of Wexford estuary there is a rich blue

P

marl full of shells, under from 18 to 30 inches of a ferriferous clayey accumulation. The former, when turned up to the surface during the drainage operations, was found to be most productive"; while the latter would not grow any crops until it was well limed. Some of the iron in this accumulation was probably due to the decomposition of the blackish, ferriferous limestone of the neighbourhood, but a great deal of it evidently came down in the waters of the Slaney and its tributaries from the Silurian and Metamorphic rock and Granite country on the west and north-west.[1]

When a surface is exposed to meteoric influences, it dries and cracks; chemical action is also at work, the latter being augmented by the frequent changes from wet to dry. At first the growth of soil goes on very slowly, as wind or rain may carry away a large portion of the disintegrated matter; once, however, vegetation begins, it gets on more rapidly, the atmospheric influences being enabled to work more readily in depth along the holes, crevices, and the like, due to the roots of the plants and the animals

[1] In general, between the blue marl and the ferruginous, clayey stuff there is a greater or less stratum nearly altogether composed of shells, as if life had suddenly been destroyed. If this has been the case, it points to a sudden partial upheaval of the land, or, as appears to me more probable, a ponding up of the water by a change in the position of the banks and shoals, the ponded-up water evaporating more or less every low tide, and the iron from it destroying life. Or fresh water may have been ponded in, which would have the same result.

associated with them. It would appear, however, that in these countries where soil has formed and a protecting envelope of grass or the like has grown on it, the increase downward is imperceptible, if any at all takes place ; the increase that takes place being upward from the vegetable growth, always assisted by the work of worms, ants, and other earth-boring animals. We are aware that this opinion is controverted, as Darwin seems to be of opinion that the formation of soil is solely due to worm-work;[1] but after careful examination and consideration, we are forced to come to conclusions different from those of that eminent observer. A foreign substance placed on grassland will be gradually covered up by a growth of soil over it. This soil Darwin seems to believe to be entirely due to the labour of earthworms, which excavate in the ground under the foreign substances and deposit over it ; so that the total thickness of the soil is not increased upwards by mould formed from vegetable decay; but all is taken from below the foreign substance and placed above it ; thereby adding to the thickness of the upper stratum of the mould, and diminishing the thickness of the portion below the foreign substance. To earthworms cannot be due the formation of all the soil, for this reason— all worm-formed mould must be ejected by them ;

[1] "Transactions of the Geological Society of London," 2d Series, vol. v. p. 505.

consequently, if the mould was solely due to them, there would be between it and the subsoil a layer of coarse sand or gravel, formed from the large particles they could not swallow. Even from Darwin's examples it would appear evident that the growth of soil cannot be due to worms alone; for spread a layer of lime on a field, and the worm, to quote that author, " is unable to swallow coarse particles, and the finer earth lying beneath would be removed by a slow process to the surface." Thus, eventually, all the matter that could be reduced small enough to be swallowed by the worms would be brought above the lime, and only the pebbles and fragments of stones left below; so that above there ought to be only this fine earth, while below there ought to be only gravel and sand. This is the result that ought to occur if to the worms alone is due the vegetable soil; but if they worked in conjunction with the decay of the vegetables, and principally in the soil due to that decay, there would be a continual shifting of a soil in which few pebbles ever existed. This, however, would involve the surface of the subsoil remaining permanent.

Nearly all the examples put forward by Darwin were observed in rich, highly-cultivated ground, where earthworms abound; therefore the growth of the worm-formed soil must have been more rapid than would ordinarily be the case, and the part added, through the decay of the vegetable matter, may not have been very apparent. Nevertheless, he seems to prove that, in one instance, the vegetable decay, not the earthworms, buried the foreign substances, namely, that of the boggy field which was covered over with a coat of gravel, and in two years and a half afterwards there was a peaty layer three-fourths of an inch thick grown over it. Very similar instances occur, and may be examined in many places in Ireland among the reclaimed cutaway bogs, or as they are locally called, *Moors*. These moors are generally tilled for a few years previous to being laid down in grass, after which a coat of marl or gravel is spread on them. If they are to be kept in good heart, more gravel or marl must be applied to them from time to time, and the drainage attended to; but if they are neglected, as is too often the case, they will tend to return to their former state, and in a short time a layer of peaty soil will grow on the marl or gravel. It is not uncommon in these moors, if a section is opened through them, to see one, two, or even three of these layers of foreign matter, pointing out the number of times the moor was "brought in," and afterwards allowed to run wild.

Any one acquainted with bogs well knows that earthworms cannot live in them; a few may be found in reclaimed moors in the made soil, both while they are in tillage and under grass; but once the original boggy nature again predominates, they disappear, so that they cannot assist in making the upper stratum of soil above the gravel; moreover, if this upper stratum is examined, it will be found to be of the same platy or rudely laminated structure, as that of all bogs that grow from the successive layers of decayed vegetable matter.

Those conversant with highly-cultivated, rich grass-land, can scarcely have failed to remark the enormous worm-work that yearly goes on; but this is not the case in all lands, for in many (not all) over chalk and limestone there is little or no worm-work; so also in poor sandy soils, or in "slob-land" newly reclaimed from the sea. On many chalk and limestone lands " stones grow," or in other words, the surface of some grass-land gradually becomes covered with stones; but in poor sandy soils, or in slob-land, although it may be the work of time, yet eventually there will be a surface-soil formed by the decay of the vegetable matter, and all the stones will be covered up by the growth of the soil. In such soils, at the beginning of the growth of the vegetable clothing, earthworms will be rare—in fact, they cannot live without organic food; therefore until the vegetable life begins they cannot exist. As the vegetable soil increases, so will the earthworms, showing that the two agents work together, also that the growth of the soil is due, not only to vegetable decay, but also to the worm-work. When the soil becomes rich, the earthworms do a greater amount of work, but while it remains poor they can do little or none, owing to the paucity of their number; so that to vegetable decay is principally due the growth of the first mould.

Lands with a permanent turf or sod, that has remained untilled for ages, may be used either as pasture or as meadow land. If the surface-soil were due only to the earthworms, then in a field all of which has the same fertility, subsoil, &c., if a part is used as permanent pasture, while from the rest hay is always cut, the condition of the whole *ought to be uniform*, and over all there would be a gradual increase of the mould. This, however, is well known to the farmer not to be the case; for the mould will increase on the pasture-land, but not on the permanent meadow, unless the latter is cut early enough to allow of a second growth or "after-grass," which is left to rot on the ground. And in corroboration of this principle, all farmers state that machine-mowing is more severe on meadow-land than hand-mowing, because the latter does not or cannot cut as close as the former; for similar reasons horses are harder on grass-land than either sheep or cows. To counteract the injurious effect of removing from the meadow-land all the vegetation which would naturally decay on it, the land has to be topdressed with foreign substances. It may be said that on pasture-land the vegetable products are not allowed to decay. This is partly true; but if such products are eaten off by the cattle, they are returned again to the surface; their fertilising quality being increased from their having been used as animal food.

In a comparison between meadows and pasture-land the difference between sole worm-work and the conjoint effects of vegetation decay and of the earth-worms appears evident. Examine a meadow-field after the hay is cut, and the worm-casts are found to be few and far between; but if the after-grass is allowed to rot on the land, the worms will be more numerous, working among the decayed vegetable matter. In pasture-land there will be a hundred worms for every one in a meadow; the greater number being found under and associated with the decayed vegetable matter in the droppings from the cattle.

In some soils, such as alluvial loam, earthworms will burrow to a great depth; but they seem usually incapable of penetrating into the ordinary subsoils that occur in Ireland. The gravelly subsoils formed by the esker-drift ought to be soft enough for them to work into, yet they are never found in it. The usual subsoil, boulder-clay drift, they never burrow into, and a not uncommon subsoil in some places, namely, a stratum of bog iron-ore, they could not possibly enter; yet over all those different kinds of subsoil the surface-mould increases if the land is laid down in permanent grass. The beginning of the formation of the soil over a subsoil of bog iron-ore cannot be due to anything but chemical action and vegetable decay, as no worm-work could have been done there; and if in such an instance a surface-soil can be formed without the aid of worms, why should not similar work go on in other places? In such places the formation of soil would be carried on under most adverse circumstances; for a bare surface of limonite is a very unfavourable place for vegetation, and usually it appears due to the decay of the water lodged on the surface, with perhaps a slight disintegration of the underlying mineral, from which are generated lichens and mosses, and eventually a peaty soil. It is added to, however, very gradually; for in dry weather, there being no depth of earth, the vegetation withers away.

Land that has been tilled to one depth for many years will have a surface to its subsoil like a road; and this is so well known, that to counteract it subsoiling has been introduced. If such land is laid down in grass, and subsequently again broken up, the surface of the subsoil caused by the former tilling will be found intact; but the depth of the mould will have increased in proportion to the number of years it has remained under grass. This increase cannot have taken place below, as the old subsoil surface still remains ; therefore the mould must have increased in thickness upwards, and necessarily by the decay of the vegetable substances. It may be said that this is an exceptional case; for this road-like surface would not be formed naturally over the subsoil.   However, if the thickness of the mould increases downward, as suggested by Darwin, is it not remarkable that over each different kind of subsoil the thickness of the vegetable mould should be so uniform?   Naturally the mould is only a few inches thick over gravel, a little thicker over clay, a good depth over a subsoil formed of a combination of clay and sand, more especially if it is limey, while over an alluvial subsoil it may be deeper still ; unless, indeed, that meteoric abrasion in each several case has removed from the surface an exact equivalent to the increase ; but this is highly improbable.

Mr John Edward Lee, F.S.A., F.G.S., the translator of Dr Keller's " Lake Dwellings of Switzerland," in a note on the depth of soil covering the " Mainland Settlement of Ebersberg," records a remarkable growth of soil at Caerleon, South Wales.   Mr Lee thus writes :—" In a field which forms the south-west portion of the ancient city of Isca Silurum, I have frequently excavated for the sake of archæology, and in one instance, when the summer was dry, the grass showed where walls were probably to be found, indicating the ancient Roman houses forming the corner

of the inhabited portion, an open space of the street
being between them and the city walls.   It was,
however, very singular that these indications, though
correct, were not verified till the ground had been
excavated to a depth of five or six feet; and before
the actual base of the wall and the floor of the street
were reached, a tall labourer was entirely hidden from
view.   The first foot or two may be accounted for by
the rubbish when the place was destroyed, but as no
soil is likely here to have been brought down by floods,
we are almost obliged to attribute the remaining
four or five feet to the annual addition of vegetable
mould. The excavation above referred to is not the only
fact which indicates a great accumulation of soil; a
handsome tesselated pavement lately discovered in
the churchyard was four or five feet below the present
surface," &c., &c.[1]   This is a startling fact in favour
of the growth of soil by vegetable decay; and Mr
Lee further states, in a subsequent communication
on the subject : " The old town of Isca is on a tongue
of land slightly above the flats on the rivers' banks
(the Usk and the Avon Llwydd), and from the situation there is no chance whatever of any earth being
brought down from the hills to the place I mentioned."
Although the growth is so large in the aggregate, yet
the yearly amount when calculated seems moderate
enough.   It must be about 1700 years since the

[1] Keller's "Lake Dwellings," p. 367.

Romans occupied the site, and if two feet are allowed for the depth of the debris of the building, there will be four feet to be formed by the growth of the soil: this is equal to 2·82 inches in a century, or ·028 inch annual growth, scarcely equal to the thickness of five sheets of foolscap paper,—or, to put it differently, one inch growth in every 35·4 years. Furthermore, all may not be due to the vegetable decay; for as the field is near Caerleon, it is probable that it has often been " topdressed," a process which would quickly add to the depth of the soil. It may be here mentioned, that some years since, a field in the north part of the County Tipperary was broken up that had a soil fourteen inches deep, and this field was known by a man seventy years old to be exactly 110 years in grass. As the average depth of soil in that part of the county is from ten to twelve inches, it would leave about 2·5 inches of soil to grow in the century; which is very similar to the results obtained near Caerleon.

Some soils are ready made when they are deposited on the surface; as, for instance, the particles carried down by flood and deposited on callows, corcasses, and other alluvial flats. Another instance is windborne soil, clouds of fine particles being carried during dry windy weather from high to low land, especially if the former is in tillage. Even grass-land may supply such materials as we have remarked

in such places as Devonshire, where after dry
weather clouds of dust are often raised off the pas-
tures. Wynne, in his description of the geology of
Kutch, India, when describing the alluvium, proves
the soil-producing powers of wind. In the early
part of the memoir he mentions the extreme
denudation the rocks suffer from the winds, and
writing of the alluvium, he states :—" The deposits
frequently resemble those of rivers ; but there are
no large rivers in Kutch, and the small intermittent
streams which it possesses would have to wander
laterally to a great extent in order to cover the
country with such detritus. They only appear to
have changed their courses in a few localities near
the coast, and their valleys, though small, are
generally single and separate, from the hills to the
sea."

Some of the Kutch alluvial deposits are salt,
sometimes two or three feet in thickness, due to
the evaporation of floods of saline water ; others are
so impregnated with salt (*Kara* and *Kuller*) as to be
incapable of producing vegetation ; while on others
(*Laana*) there is a scanty growth of plants.

Other soil-formers that are rarely thought of are
the ants. These workers, although such pigmies in this
country, and therefore, compared with the earthworms,
less capable, individually, of work, are so numerous
and energetic, that in the special places to which

they resort their yearly work is much more conspicuous than the annual worm-work. In Leicester-
shire, for instance, and elsewhere in central England, the hill-building ants are a nuisance to the
grazier.   In the spring the ants may be observed
beginning their work, and from that time they
carry on their operations all through the summer,
to the late autumn or early winter.   On a moor
they raise hillocks scattered about over the ground,
getting closer and closer together as the colonies
increase.   But where the ants are most useful is in
places where crags and large stones are mixed up
with patches of sandy peat.   In such a locality they
will always build on a rock, the foundation of their
habitation being at its junction with the soil.   These
ants seldom raise their structures as high as the
moorland ants do, but they make up in extent, for
deficiency in height; and by this means the rocks
are gradually covered with soil and vegetation; for,
on account of the season of the year at which they
build, the growth of the plants keeps pace with the
formation of the ant-soil, and protects it in a great
measure from meteoric abrasion.   During the winter,
however, the shape of the ant-hills is somewhat
modified; but the matted roots of the plants preserve
the major part of the soil, which will thus remain,
forming an envelope for the rock.

As the annual work of one colony is rarely less

than two square feet in superficial area, and sometimes exceeds a square yard, the yearly work done by a number of colonies must be very considerable. The ants that build in the sandy moors seem to frequent the same spot year after year; and some of the hillocks, after a number of years, are of a considerable height; some that were measured in the County Cork being four feet high, and nearly a yard in diameter at the base. The ants that build on rocks appear to occupy generally new places every year; but perhaps not always, as it has been observed that in some places the new work has been carried on in a place where other works had existed in former years. The rock-covering ants are undoubtedly the most useful animals of the two; but the others have also a serviceable place in nature, as they change a cold unprofitable peat into a workable vegetable soil.

Other soil-producing animals in these countries are moles, rabbits, rats, mice, foxes, badgers—in fact, all animals that burrow in the ground, including the "tillers of the ground." Some do a considerable quantity of work, while others only act in a small way. Moles are among the most active workers; so are rabbits; the latter often in a great measure helping to change cliffs into sloping escarpments; while foxes and badgers in general do very little work, as they will rather occupy natural holes than burrow for themselves.

Some soils may be due to the decay of animal matter. Off the coast of South America there are large deposits of guano, which are generally believed to be due to vast accumulations of the excrements of birds. It has, however, been shown by the late Dr J. R. Kinahan that the guano accumulated on the Chinchas Islands, Peru, are not the excrements of birds, but "are formed of layers of seals' dung and decayed seals, the denser, white, thin layers being made up of the former, and the more friable, darker, thick layers of the latter," as the sea-lions resort in hundreds to such places to die. At the same time this observer does not wish to state all guano is similarly formed, as the guano from Ichaboe and South Australia is so different from that of Peru that they would seem to have a different origin.[*]

[*] *Journal of the Royal Dublin Society*, July 1856, pp. 89 *et seq.*

PRINTED BY BALLANTYNE AND COMPANY
EDINBURGH AND LONDON

www.ingramcontent.com/pod-product-compliance
Lightning Source LLC
Chambersburg PA
CBHW021517210326
41599CB00012B/1292